The Student Edition of MICRO-LOGIC II™

Limited Warranty

Addison-Wesley warrants that the Student Edition of MICRO-LOGIC II ("the program") will substantially conform to the published specifications and to the documentation during the period of 90 days from the date of original purchase, provided that it is used on the computer hardware and with the operating system for which it was designed. Addison-Wesley also warrants that the magnetic media on which the program is distributed and the documentation are free from defects in materials and workmanship during the period of 90 days from the date of original purchase. Addison-Wesley will replace defective media or documentation or correct substantial program errors at no charge, provided you return the item with dated proof of purchase to Addison-Wesley within 90 days of the date of original purchase. If Addison-Wesley is unable to replace defective media or documentation or correct substantial program errors, your license fee will be refunded. These are your sole remedies for any breach of warranty.

Except as specifically provided above, Addison-Wesley makes no warranty or representation, either express or implied, with respect to this program, documentation or media, including their quality, performance, merchantability, or fitness for a particular purpose. Spectrum makes no warranty or representation, either express or implied, with respect to this program, documentation or media, including their quality, performance, merchantability or fitness for a particular purpose.

Because programs are inherently complex and may not be completely free of errors, you are advised to verify your work. **In no event will Addison-Wesley or Spectrum be liable for direct, indirect, special, incidental, or consequential damages arising out of the use of or inability to use the program, documentation or media,** even if advised of the possibility of such damages. Specifically, neither Addison-Wesley nor Spectrum is responsible for any costs including, but not limited to, those incurred as a result of lost profits or revenue (in any case, the program must be used only for educational purposes, as required by your license), loss of use of the computer program, loss of data, the costs of recovering such programs or data, the cost of any substitute program, claims by third parties, or for other similar costs. In no case shall the liability of Addison-Wesley exceed the amount of the license fee and in no case shall Spectrum have any liability.

The warranty and remedies set forth above are exclusive and in lieu of all others, oral or written, express or implied. No Addison-Wesley dealer, distributor, agent, or employee is authorized to make any modification or addition to this warranty.

Some statutes do not allow the exclusion of implied warranties; if any implied warranties are found to exist, they are hereby limited in duration to the 90-day life of the express warranties given above. Some states do not allow the exclusion or limitation of incidental or consequential damages, nor any limitation on how long implied warranties last, so these limitations may not apply to you. This warranty gives you specific legal rights and you may also have other rights which vary from state to state.

To obtain performance of this warranty, return the item with dated proof of purchase within 90 days of the purchase date to: Addison-Wesley Publishing Company, Inc., Educational Software Division, Jacob Way, Reading, MA 01867.

The Student Edition of MICRO-LOGIC II™

A digital circuit simulation program . . . *adapted for education*

Lee D. Coraor

The Pennsylvania State University

Addison-Wesley Publishing Company, Inc.

Benjamin/Cummings Publishing Company, Inc.

Reading, Massachusetts • Redwood City, California • New York
Don Mills, Ontario • Wokinham, England • Amsterdam • Bonn
Sydney • Singapore • Tokyo • Madrid • San Juan • Milan • Paris

The Student Edition of MICRO-LOGIC II is published by Addison-Wesley Publishing Company, Inc. and Benjamin/Cummings Publishing Company, Inc. Contributors included:

Alan Jacobs, Executive Editor
Dana Degenhardt and Holly Wallace, Product Managers
Carol Botteron, Developmental Editor
Stephanie Kaylin, Copyeditor
Mary Coffey, Senior Production Supervisor
Karen Wernholm, Software Production Supervisor
Lu Anne Piskadlo, Manufacturing Media Supervisor
Marshall Henrichs, Corporate Art Director
Jean Seal, Cover Design

The Student Edition of MICRO-LOGIC II was developed and programmed by Spectrum Software, Inc.

0-201-50554-1
0-201-50552-5 (5 1/4" disk format)
0-201-50511-8 (3 1/2" disk format)

5678910-AL-9594939291

Preface

Digital logic simulation, or logic simulation, is a state-of-the-art design methodology that allows for the rapid testing of prototypes of alternative circuit designs; operation of a logic circuit can be studied without having to build the circuit.

A logic simulator such as MICRO-LOGIC II has dual functions: as a Computer-Aided Design drafting tool, and as a logic simulation tool. When you use the simulator as a drafting tool, you can think of the CRT screen as a large sheet of drafting paper on which various logic circuits can be drawn. The simulator component of the package simulates the operation of a logic circuit. Specific input signals are defined that are applied to the circuit inputs. The logic simulator then simulates the operation of the circuit and generates the circuit outputs based on the defined inputs.

Using This Manual

This manual was designed primarily for the non-mouse user. Most of the functions of MICRO-LOGIC II can be implemented without a mouse, and the first seven tutorials are written as if a mouse is not available. However, special sections dealing with mouse commands are included as appropriate. These sections update features for the mouse user. It is suggested that the mouse user preview these sections and then proceed through the tutorial making the appropriate substitutions. Additional mouse commands are presented in Tutorial 8.

In each of the eight tutorials, more basic and/or commonly used information and procedures are covered first, and then expanded on in later tutorials.

Tutorial 1 familiarizes the user with the MICRO-LOGIC II program basics. A design example, a full adder circuit, is introduced that is followed through the first four tutorials. This tutorial introduces basic drawing principles and file-handling procedures.

Tutorial 2 continues the design of the full adder circuit by including text (labels) and lines.

Tutorial 3 introduces the use of data channels (inputs) and the generation of reports, helpful in debugging drawings.

Tutorial 4 completes the design example of the full adder circuit by simulating the performance of the circuit. At this point the student can simulate any combinational circuit.

Tutorial 5 introduces sequential components and the use of clock signals in a circuit. A second design example, a synchronous binary counter, is introduced.

Tutorial 6 performs the simulation of the synchronous binary counter drawn in the previous tutorial. Variables in simulation generation, component characteristics, and propagation delay are introduced.

Tutorial 7 introduces macro drawings, the creation of new components and macros, wired-OR and wired-NOR connections, and tri-state devices.

Tutorial 8 covers features that are available only with a mouse.

Also provided are a Reference Section, which contains a Quick Reference Guide, and a Glossary.

Differences Between the Professional and Student Versions

The professional version of MICRO-LOGIC II can be differentiated from the student version in the following ways:

- The professional version includes a Shape editor.
- It also includes a plotter interface.
- It allows for the results of simulation runs to be saved to and retrieved from disk.
- It includes an expanded library of TTL and CMOS components.
- It permits the use of an EGA card and monitor.
- It has the following maximum limits:

Components in a drawing	1000
Lines in a drawing	4000
Text labels in a drawing	200
Data channels	32
Clocks	10
Nodes or gates after macro expansion	10000
Drawing files on disk	300
Library shapes	75

- It requires a 640K system with a hard disk.

Maximum Limits in the Student Edition

Components in a drawing	500
Lines in a drawing	1000
Text labels in a drawing	200
Data channels	32 (16 predefined)
Clocks	10 (5 predefined)
Nodes or gates after macro expansion	500
Drawing files on disk	100

Contents

I Getting Started

1 Before You Begin

This part of the manual describes the contents of the Student Edition of MICRO-LOGIC II and the procedures for installing, starting, and ending MICRO-LOGIC II.

Checking Your MICRO-LOGIC II Package

Your package for the Student Edition of MICRO-LOGIC II should contain the following items:

- The user's manual (this book)
- One program disk and one data disk, both in standard 5.25-inch 360K format
  or
- One 3.5-inch 720K disk containing both program and data files

Checking Your Computer Setup

Your personal computer system should be compatible with the following MICRO-LOGIC II hardware requirements:

- An IBM PC, XT, AT, or PS/2 compatible computer
- An operating system of 1.2 megabytes, MS-DOS version 3.0 or higher
- One of the following graphics adapters, or its equivalent:
    - IBM Color Graphics Card (CGA) and monitor
    - Hercules monographics adapter and IBM monochrome display
- At least 512K of RAM for a two disk drive system, or 640K for a hard disk drive system
- Epson-compatible graphics, although a printer is optional
- A Microsoft or compatible mouse (highly recommended, although the system will run without it)

Product Support

Telephone assistance is provided to registered instructors who have adopted the Student Edition of MICRO-LOGIC II. However, neither Addison-Wesley nor Spectrum provides telephone assistance to students.

If you encounter difficulty using the Student Edition software, read the section of this manual containing the information on the commands or procedures you are trying to execute.

If you must ask your instructor for assistance, describe your question or problem in detail. List the steps or procedures you followed when you encountered the problem so that your instructor may best understand how to assist you. Include any error messages, written exactly as they appeared on the screen.

Typographical Conventions

You control MICRO-LOGIC II through the keyboard and/or mouse of your personal computer. In this manual, special symbols and a special, computer-like typeface tell you what keystrokes to enter. For example:

Press: [↵]

Type: `L`

Type: `EXOR`

Symbols for special keys include the following:

[↵]	Carriage return (enter key)
[↑]	Up cursor
[↓]	Down cursor
[←]	Left cursor
[→]	Right cursor
[Ins]	Insert key (below cursor keys)
[←]	Backspace key (top row, with arrow pointing left)
[Del]	Delete key (below cursor keys)
[Esc]	Escape key (top row, left side)
[⇆]	Tab key (at left; may have arrows pointing left and right)
[PgUp]	Page up key (with cursor keys)
[PgDn]	Page down key (with cursor keys)
[Home]	Home key (with cursor keys)
[End]	End key (with cursor keys)
[Scroll Lock]	Scroll key (top right)
[Alt]	Alternate key (bottom left)
[Ctrl]	Control key

[+] + key (at right, not in top row)

[−] - key (above + key)

Special characters include:

μ Lowercase Greek mu

⊕ Exclusive Or operator (+ in a circle)

• And operator (raised dot)

Note that although "Press" and "Type" both mean to strike or press a key, each has a special meaning:

- "Press," used primarily with special keys, cursor keys, and the [↵] key, performs an operation. For example, if you are to press the delete key, the text will say

  Press: [Del]

- "Type," used primarily with standard typewriter keys, enters information. The keys to be typed are either in **boldface** type or in a special computer-like type. For example, if you are to type the word **start**, the text will say

  Type: `start`

The Shift, Control (Ctrl), and Alternate (Alt) keys are used only in combination with other keys; you hold down one of these special keys while striking another key. Such combinations are shown in this book using their special key symbols, for example, [Alt] `F`.

2 Setting Up MICRO-LOGIC II

Setup procedures depend on the kind of system you have. This chapter gives procedures for a system with two floppy disk drives and for a system with a hard disk.

Note to 3.5-inch Disk Users: If you are using the 3.5-inch disk version of MICRO-LOGIC II, the disk will contain both program and data files (which are on separate disks for the 5.25-inch disk version). You will need to follow the procedure for copying disks only once—both the program files and the data files will be copied. Whenever the instructions call for switching to the data disk, you can ignore the procedure that follows, since you already have those files on your 3.5-inch disk.

Floppy Disk Drive System

You need the following items to install MICRO-LOGIC II:

- The MICRO-LOGIC II program disk and data disk
- The MS-DOS disk
- One or more blank disks with sleeves and labels

Use DOS to make a backup copy of the MICRO-LOGIC II program disk, following the directions in this section.

Starting Your Computer

With your computer turned off, put the DOS disk in drive A and close the door.

Turn the computer on, and wait for it either to display the date and time or to ask you for this information. If requested to do so, type in the date as MM-DD-YY and press ↵. For example, to enter the date July 15, 1989,

Type: `07-15-89`

Press ↵

If requested to do so, type in the time as HH:MM:SS, using a 24-hour clock. For example, to enter the time 1:45 p.m.,

Type: `13:45:00`

Press: ↵

If you do not enter the date and time in the proper form, DOS will let you try again. Once the date and time are correct, the system prompt appears (usually as **A>**). You are now ready to format new disks.

Formatting a New Disk

You must format two disks to use as your working MICRO-LOGIC II program disk and data disk. When you format a disk, you lose any data previously stored on it.

With the DOS disk in drive A,

Type: `FORMAT B:`

Press: ↵

Your screen should say **Insert new diskette for drive B: and strike any key when ready.**

Put a blank diskette in drive B and close the door. To start formatting,

Press: ↵

The message **Formatting ...** appears on the screen. You can see the light next to the drive door go on, and you can hear the drive. Formatting can take as long as one minute. When the procedure is finished, the light goes out and the message **Format complete** appears. DOS then displays some information about the disk and asks, **Format another (Y/N)?**

If you want to format an additional disk, type **Y** and press ↵ if necessary.

Remove the formatted disk from drive B, label the disk to identify it, and return it to its sleeve. Repeat the formatting procedure for as many disks as you'd like to format. When you are finished formatting disks, type **N** and press ↵ when asked, **Format another (Y/N)?** The **A>** prompt will appear.

Making Backup Disks

Put the MICRO-LOGIC II program disk in drive A and close the door. Put a formatted blank disk in drive B and close the door. At the DOS prompt,

Type: `COPY *.* B:`

(Be sure to leave a space before the B.)

Press: ↵

DOS will now copy the files from the program disk in drive A to the formatted blank disk in drive B. The asterisk stands for any filename or file extension. Thus the ***.*** tells DOS to copy every file. As it copies each file, DOS displays the name of that file. When finished, DOS displays the number of files it has copied, and the **A>** prompt appears.

If you are using the 3.5-inch disks this procedure will copy both the program files and the data files.

Now remove the program disk from drive A and the backup disk from drive B. Label the backup disk, using the name of the original disk and indicating that this is a backup copy. Copy the data disk in the same way. Leave the data disk in drive A to unpack the files.

As you use MICRO-LOGIC II, drawings you save will be stored on the data disk. To avoid losing a drawing if the disk is damaged, copy the data disk every time you save a drawing.

Unpacking Files

The files for the macros must be unpacked before they can be used. (You'll learn about macros in detail in Tutorial 7, but you'll use them earlier.) To unpack the files, be sure the original data disk is in drive A and the backup data disk is in drive B. At the **A>** prompt,

Type: `UNPACK A: B:`

Press: ↵

The screen will display the message **Unpacking**, the name of each file as it is unpacked, and finally the **A>** prompt.

Store the original disks in a safe place, always using the backup copies to run MICRO-LOGIC II. This way, if anything happens to your backup copies, you can make new backup copies using the originals.

Erasing Unneeded Files

Your copy of the data disk contains files that you no longer need: the unpacking directions and the compressed files that have been unpacked. Their filenames are as follows:

GATES.TMP
LINEH.TMP
LINEV.TMP
STRINGS.TMP
KEY.TMP
UNPACK.EXE

To save space on the disk, you can erase these files using the command **erase** and the filename. To erase the first unneeded file, at the **A>** prompt,

Type: `ERASE GATES.TMP`

Press: ↵

When the file is erased, another **A>** prompt will appear. Erase the other files in the same way.

Once you have completed the tutorials, you may wish to delete some of the files from the data disk in order to provide more room for your own data files. Chapter 7 of the Reference Section includes a list of the files that must be kept for the program to work.

Mouse Software

If you are using a mouse, you must also install the mouse software on your working copy of MICRO-LOGIC II. To do so, put the MS-DOS or compatible software disk in disk drive A and close the door. Put your backup MICRO-LOGIC II disk in drive B and close the door. To make sure the DOS prompt **A>** is on the screen,

Type: `A:`

Press: ↵

Type: `A:MOUSE.COM B:`

Press: ↵

(If the instructions in the manual that came with your mouse are different, use them instead.)

DOS will now copy the mouse software to the program disk.

Hard Disk Drive System

If you have a hard disk drive system, you will probably want to copy MICRO-LOGIC II onto the hard disk. Once the software is installed, you can start and run MICRO-LOGIC II and store the documents you create without having to handle floppy disks.

These instructions assume that you have DOS on your hard disk and that your hard disk drive is designated C. If this is not the case (or if you're not sure), ask your dealer or the technical resource person at your school for assistance.

If your computer is on and displays the **C>** prompt, you can skip the following section, Starting Your Computer.

Starting Your Computer

Turn the computer on and wait for it either to display the date and time or to ask you for them.

If requested to do so, type in the date in the form MM-DD-YY, and press ↵. For example, if the date is July 15, 1989,

Type: `07-15-89`

Press: ↵

If requested, type in the time in the form HH:MM:SS, using a 24-hour clock. For example, if the time is 1:45 p.m.,

Type: `13:45:00`

Press: ↵

If you enter the date or time in the wrong form, DOS will let you try again. Once the date and time are correct, a **C>** prompt appears. If an **A>** prompt appears instead, type **c**: and press ↵ . A **C>** prompt will appear. You are now ready to create a subdirectory for MICRO-LOGIC II.

Creating a Subdirectory for MICRO-LOGIC II

Directories let you organize the files on your hard disk into groups. If MICRO-LOGIC II is to run properly, the main system files must be in the same directory. It is a good idea to have a subdirectory that contains only the MICRO-LOGIC II files. You will probably find it convenient to keep your MICRO-LOGIC II documents in the same directory as the program, although you can put them elsewhere.

At the **C>** prompt,

Type: `MD C:\ML2`

Press: ↵

Type: `CD ML2`

Press: ↵

Copying Disks

To copy the MICRO-LOGIC II files onto your hard disk, insert the program disk into the floppy disk drive and close the disk drive door. (If you have a 3.5-inch disk, insert the program/data disk into the disk drive.)

Type: `COPY A:*.* C:`

(Be sure there is a space before the C.)

Press: ↵

An asterisk stands for any filename or file extension. Thus the ***.*** tells DOS to copy every file. The file is displayed as it is copied. After all the files are copied, DOS displays the number of files copied, and a **C>** prompt appears.

If you are using 5.25-inch disks, copy the data disk in the same way. Leave it in the floppy disk drive to unpack the files, following the directions in the next section.

If you are using a 3.5-inch disk, this procedure has copied both the program and the data files. Leave the disk in the floppy disk drive to unpack the files, following the directions in the next section.

Unpacking Files

The files for the macros must be unpacked before they can be used. (You'll learn about macros in detail in Tutorial 7, but you'll use them earlier.) To unpack the files, put the data disk in drive A. At the **C>** prompt,

Type: `UNPACK A: C:\ML2`

Press: ↵

The screen will display the message **Unpacking**, the name of each file as it is unpacked, and finally a **C>** prompt.

Store the original disks in a safe place. This way, if anything happens to your hard disk, you can recopy the files using the originals.

Erasing Unneeded Files

Your copy of the data disk contains files that you no longer need: the unpacking directions and the compressed files that have been unpacked. Their filenames are as follows:

GATES.TMP
LINEH.TMP
LINEV.TMP
STRINGS.TMP
KEY.TMP
UNPACK.EXE

To save space on the disk, you can erase these files using the command **erase** and the filename. To erase the first unneeded file, at the **C>** prompt,

Type: `ERASE GATES.TMP`

Press: ↵

After the file is erased, another **C>** prompt appears. Erase the other files in the same way.

Once you have completed the tutorials, you may wish to delete some of the files from the data disk in order to provide more room for your own data files. Chapter 7 of the Reference Section includes a list of the files that must be kept for the program to work.

Mouse Software

If you are using a mouse, you must also install the mouse software on your hard disk. To do so, put the MS-DOS or compatible software disk in disk drive A and close the door. To make sure the DOS prompt **A>** is on the screen,

Type: `A:`

Press: ↵

Type: `COPY A:MOUSE.COM C:\ML2`

Press: ↵

(If the instructions that came with your mouse are different, use them instead.)

DOS will now install the mouse software on your hard disk in subdirectory ML2. The prompt **C>** will appear when the installation is complete.

If the mouse software is already loaded on your hard disk, you should remove the line that says **mouse** from your start.bat file.

Making Backup Disks

It is a wise precaution to make backup copies of any drawings and other new files that you save. First you must format a floppy disk, following the instructions earlier in this chapter. When you format a disk, you lose any data previously stored on it.

To copy files onto a formatted blank disk, put the disk in the drive and close the door. To make sure that you are in the ML2 directory,

Type: `DIR`

Press: ↵

To stop the directory from scrolling, press Ctrl **S.** While the directory is stopped, you can write down the exact names of the files you want to copy. (A filename consists of the letters and numbers in the first column of the directory, followed by a period and the three letters in the second column—for example, **BOX.DWG.**)

To resume scrolling, press Ctrl **S** again.

To copy one file onto the floppy disk,

Type: `COPY FILENAME A:`

Press: ↵

To copy all the **.DWG** files,

Type: `COPY *.DWG A:`

Press: ↵

DOS will display the names of the files as it copies them. When finished, DOS will display the number of files it has copied, and the system prompt will appear.

You can copy all the **.PAT** files or all the **.MAC** files in the same way.

Now remove the backup disk and label it.

Starting the Program

Once you have installed MICRO-LOGIC II as either a working copy of the program floppy disk or as a subdirectory on your hard disk, you can start up MICRO-LOGIC II and use it as many times as you wish.

The start-up procedures vary according to the type of storage equipment you are using.

Running MICRO-LOGIC II from a Floppy Disk

If your computer is turned off, insert the DOS disk in drive A, close the door, and turn on the computer. Provide date and time information if requested. Then remove the DOS disk and load the copy of the program disk in drive A.

If your computer is on and the screen displays the DOS prompt **A>**, insert the program disk in drive A and the data disk in drive B.

The next step depends on your display equipment. If you have an IBM Color Graphics Adapter (CGA),

Type: `START`

Press: ↵

If you have a Hercules monographics adapter,

Type: `START H`

Press: ↵

If you are restarting MICRO-LOGIC II but you have not turned off the computer or rebooted it since you last used MICRO-LOGIC II,

Type: `START`

Press: ↵

If you are using mouse software for the first time, you must initialize it. With either type of display equipment, if you are using mouse software other than Microsoft, initialize it by typing in the appropriate replacement call. If you are using Microsoft, the MOUSE.COM file is called up automatically. If you don't have a Microsoft or compatible mouse, you'll see the message **Bad command or file name**.

You will see a title screen, and then the MICRO-LOGIC II Designer Display screen. You are now ready to use MICRO-LOGIC II.

The title and copyright notice may remain on the screen and the message **Insert data disk and enter drive letter (A, B, C...Z)** may appear. If this occurs, insert the data disk into drive B and

Type: `B`

Press: ↵

Running MICRO-LOGIC II from a Hard Disk

Turn on your computer and call up your MICRO-LOGIC II subdirectory.

Type: `CD ML2`

Press: ↵

The next step depends on your display equipment. If you have an IBM Color Graphics Adapter (CGA),

Type: `START`

Press: ↵

If you have a Hercules monographics adapter,

Type: `START H`

Press: ↵

If you're restarting MICRO-LOGIC II but you haven't turned off the computer or rebooted it since you last used MICRO-LOGIC II,

Type: `START`

Press: ↵

If you are using mouse software for the first time, you must initialize it. With either type of display equipment, if you are using mouse software other than Microsoft, initialize it by typing in the appropriate replacement call. If you are using Microsoft, the MOUSE.COM file is called up automatically. If you don't have a Microsoft or compatible mouse, you'll see the message **Bad command or file name**.

You'll see a title screen, and then the MICRO-LOGIC II Designer Display screen.

The title and copyright notice may remain on the screen and the message **Insert data disk and enter drive letter (A, B, C...Z)** may appear. If you have copied the data disk to the hard drive,

Type: `C`

Press: ↵

If you haven't copied the data disk to the hard drive, insert the data disk into drive A or B.

Type: `A` or `B`

Press: ↵

Ending the Program

You may exit MICRO-LOGIC II at any time on either a floppy or hard disk drive system by pressing [Esc] until you have completely exited MICRO-LOGIC II and the DOS prompt **A>** or **C>** appears on the screen.

II Tutorial Exercises

1 Creating Logic Diagrams

1 Creating Logic Diagrams

In Tutorial 1 you'll learn how to:

- Interpret and use the Designer Display screen
- Manipulate components in a drawing
- Save, load, and delete drawings

The Designer Display Screen

After you start MICRO-LOGIC II, your screen should look like Figure 1.1, the **Designer Display screen**. The Designer Display screen is the large "drafting paper" on which the logic diagram will be drawn. From this screen some basic housekeeping tasks can also be performed—for example, saving, loading, and deleting drawings. In addition, other functions of MICRO-LOGIC II, including the simulation of a circuit, can be executed from the Designer Display screen.

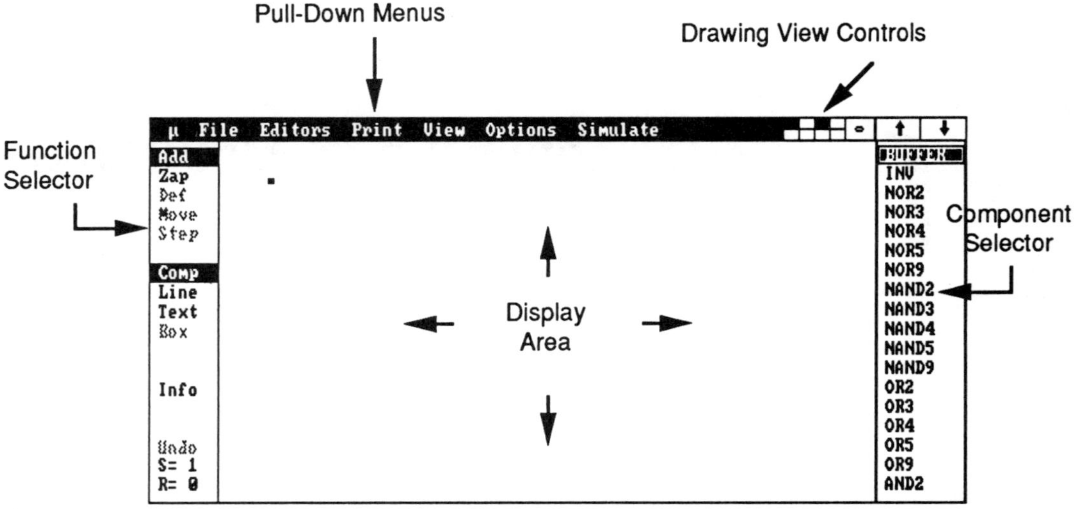

Figure 1.1

As shown in Figure 1.1, the Designer Display screen consists of five major areas:

- Pull-down Menus
- Drawing View Controls
- Function Selector
- Component Selector
- Display Area

Brief overviews of each area are presented here.

Additional details are provided in the remaining tutorials as needed.

Pull-Down Menus

Seven pull-down menus run across the top left of the screen:

- µ
- File
- Editors
- Print
- View
- Options
- Simulate

To pull down a menu, press [Alt] along with the first letter of the menu name. For example, to use the μ menu,

Press: [Alt] U

Mouse To access a menu, move the mouse pointer to the menu title, and then hold down the button and drag. Slide the mouse pointer to the desired menu option—for example, the μ—and release the button.

The μ menu has now been pulled down as shown in Figure 1.2. This menu displays some information about the MICRO-LOGIC II program along with the time and the date. If the time and date are incorrect, correct them following instructions in your DOS manual. If you have a real-time clock installed, consult its manual to reset the time and date.

You can exit MICRO-LOGIC II from the μ menu by pressing [Esc]. To leave this menu without exiting the program, press any key except the [Esc] key.

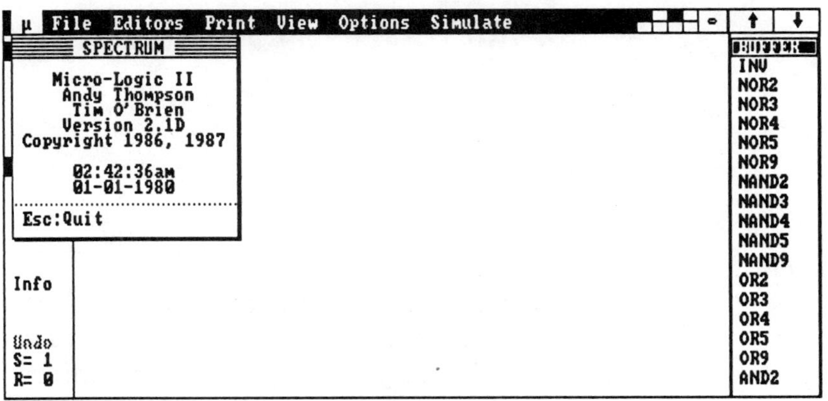

Figure 1.2

You can also exit MICRO-LOGIC II from the Designer Display screen by pressing [Esc] (a more convenient way for a keyboard user to exit the program). The μ menu may be more convenient for a mouse user.

Now pull down the **Print** menu:

Press: [Alt] P

Three options are displayed (see Figure 1.3):

- 1:Print drawing
- 2:Print reports
- 3:Plot drawing

An option can be selected by typing the appropriate number. If you do not wish to select any of these options, you can exit the menu by pressing any non-numeric key, including [Esc].

We will examine the **1:Print drawing** and **2:Print reports** options later.

The plotter option is not available in the student version of MICRO-LOGIC II, but let's attempt to use it to illustrate the use of the menus. Select option **3:Plot drawing**:

Type: 3

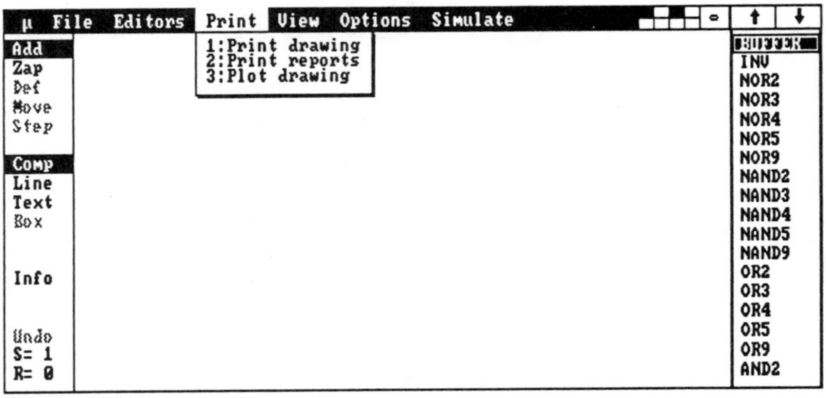

Figure 1.3

A **dialog box** has now appeared on your screen (see Figure 1.4). A dialog box provides you with necessary information and waits for your response to this information. The dialog box in Figure 1.4 tells you that the plot drawing option is not available in the student version of MICRO-LOGIC II; your response simply acknowledges this. This dialog box allows for only one response: **Ok:**

Press: [Alt] O

Some dialog boxes have more than one possible response (for

example, yes or no) and/or may require more than one response.

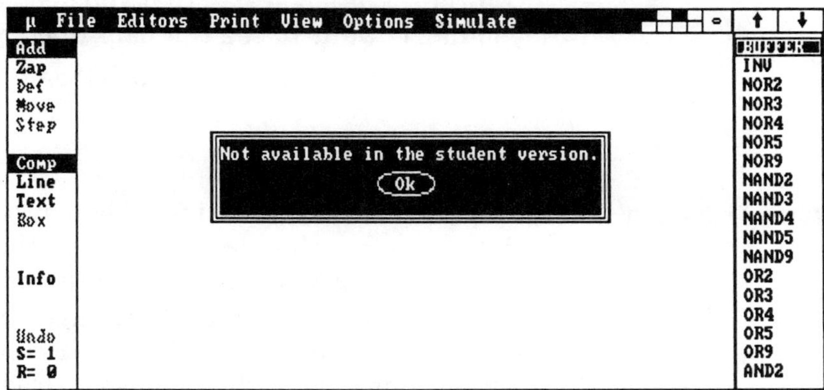

Figure 1.4

Mouse You can respond to a dialog box by moving the mouse pointer to the correct response button in the dialog box and clicking.

Now let's look at the **Options** menu. Pull down the Options menu:

Press: [Alt] O

Five options are available:

- 1:Wire-Nor
- 2:Wire-Or
- 3:Colors
- 4:Help lines
- 5:Mouse ratio

(If option **3:Colors** is shaded, you cannot select this option because of the limitations of the graphics adapter in your system.)

Select option **4:Help lines:**

Type: 4

You now see an informational help line at the bottom of the

screen. The message **Adds comp. when mouse is clicked or Ins or A key is pressed** appears. This message indicates what the currently selected functions in the Function Selector (discussed later) will do. The help line may assist you in using the tutorials; it is suggested that you keep this option activated. If you don't want to see the help lines, pull the Options menu down and type **4** again. This will toggle the help lines option off.

Drawing View Controls

A second area of the Designer Display screen, located in the top right corner of the screen, contains the **Drawing View controls**. The Drawing View controls indicate what part of the drafting paper is currently displayed on the screen. They can also be used to control the positioning of the paper on the screen, but only with a mouse. Drawing View controls are covered in Tutorial 2.

Function Selector

The **Function Selector** runs down the left side of the Designer Display screen. The Function Selector allows you to add to, zap (delete), or do something else to a component, line, or other element of the drawing. Currently selected functions are highlighted on the screen. (The Add and Comp functions are default selections whenever MICRO-LOGIC II is started.) A function is selected by typing its first letter. For example, to select the Line function,

Type: `L`

Notice that **Line** is now highlighted in the Function Selector. This indicates that a line can be added to the drawing. Now,

Type: `T`

Notice that **Text** is now highlighted in the Function Selector, indicating that text can be added to the drawing. The remaining functions in the Function Selector are discussed later.

Component Selector

The Designer Display also includes the **Component Selector,** which runs down the right side of the screen. Every component that can be used in a drawing is listed in the Component Selector. You can page through the Component Selector by pressing the tab key, ⇆. (The ⇆ key is near the left side of the keyboard, and may be labeled with two arrows pointing left and right.)

Mouse To examine the Component Selector, click either the up arrow or the down arrow just above it. To select a particular component in order to add it to a drawing, click that component on the Component Selector.

Three components are flagged by boxes around them. When you press ⇆, the Component Selector advances to the next flagged component. This makes it easier to select components you use frequently. You can flag other components by moving the mouse pointer to the desired component and clicking the right-hand button. You can unflag a component by performing the same procedure, but only when it is not the currently selected (highlighted) component.

Components can be classified into two groups. The first group, composed of **basic components,** includes individual Boolean gates, data channels, and clocks. The second group is composed of **macros,** which include flip-flops and integrated circuit packages. A macro can be used in essentially the same way as a basic component. Tutorial 7 defines macros and describes their unique characteristics.

Boolean gates consist of the following:

- Buffer
- Invertor
- Not Or (NOR)
- Not And (NAND)
- Or (OR)
- And (AND)
- Exclusive Or (EXOR)

These gates have varying numbers of inputs. They are labeled in the Component Selector according to their function and their number of inputs. For example, a 5-input NAND gate is listed as NAND5, and a 5-input OR gate as OR5. Find these components in the Component Selector.

A **data channel** is a component that represents a circuit input. Tutorial 3 defines and discusses these components.

A **clock** is a component that represents a clock signal; clocks are covered in Tutorial 5.

The **flip-flop** macros include a latch and a variety of the four basic flip-flops: J-K, R-S, T, and D. Except for the latch, which is labeled LATCH, these macros are labeled with a 2- to 4-letter abbreviation. The first letter indicates the flip-flop type (J, R, T, or D). The second letter indicates the type of triggering: N for a negative edge trigger, P for a positive edge trigger, and M for a master-slave device. Any remaining letters indicate whether the device has separate preset (P) and/or clear (C) inputs. For example, a DPPC is a D-type flip-flop that is positive edge-triggered and has both preset and clear inputs. Use the ⊟ key to page through the Component Selector until you find the DPPC component.

The **integrated circuit packages** are macros that implement standard TTL integrated circuit packages. These macros are labeled by their standard TTL component number. For example, the 7486 is a quadruple 2-input Exclusive Or package and is labeled in the Component Selector as 86. Use the ⊟ key to find the 7486 component.

Note: Only a portion of the component names in the Component Selector can be viewed by pressing ⊟ (although all of the component names can be viewed by using a mouse). For instance, none of the data channels or clocks can be viewed by pressing ⊟. With the mouse, you use the Component Selector to actually select a component. Without a mouse, you type in the name of the component instead. To remember component names, use the Component Selector or the Component editor (see Tutorial 7). Components are also listed in the Reference Section of this manual.

Display Area

The final area of the Designer Display screen is the **Display area**, in the center of the screen. The Display area represents the sheet of drafting paper on which a logic diagram is drawn. Somewhere in the Display area is a small blinking dot, or **cursor**, which defines where elements will be added or deleted in the Display area. The cursor can be moved by using the four **cursor keys:** [→], [←], [↑], and [↓]. The cursor keys are close to the right side of the keyboard and have arrows pointing right, left, up, and down.) Try using the cursor keys to move the cursor from one side of the Display area to the other and from the top to the bottom.

Full Adder Example

A full adder is a combinational circuit with three inputs and two outputs. The full adder is used to add two 1-bit numbers and a carry input, and generates a sum and a carry output. A truth table of the full adder is shown in Table 1.1, where the two inputs are IN1 and IN2, the carry input is CIN, the sum is SUM, and the carry output is COUT.

CIN	IN2	IN1	COUT	SUM
0	0	0	0	0
0	0	1	0	1
0	1	0	0	1
0	1	1	1	0
1	0	0	0	1
1	0	1	1	0
1	1	0	1	0
1	1	1	1	1

Table 1.1

Boolean expressions for the sum and the carry output are as follows:

SUM = IN1⊕ IN2 ⊕ CIN

COUT = ((IN1⊕ IN2) · CIN) + (IN · IN2)

A logic diagram for the implementation of the full adder, using the preceding equations, is shown in Figure 1.5. Let's use this example in this tutorial, as well as in Tutorials 2–4, to practice drawing logic diagrams and using the simulator. Here we'll practice adding the components to the drawing.

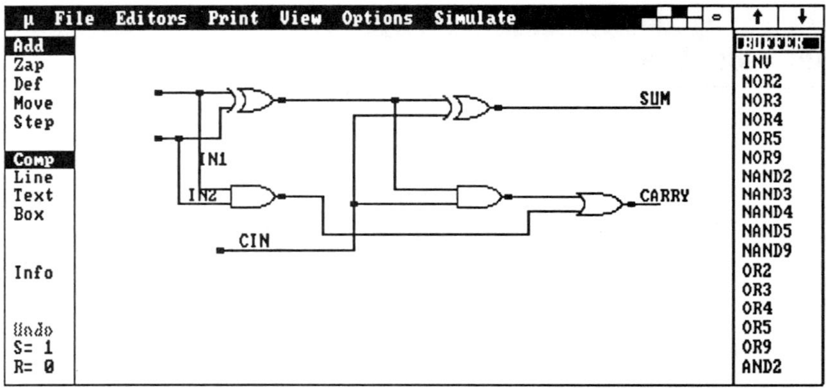

Figure 1.5

Adding Components

First let's add the left EXOR gate shown in Figure 1.5. Using the cursor keys, move the cursor to a point on the screen about a quarter of the way across the Display area from the left edge and about a quarter of the way down the Display area from the top. A component can be added here by selecting **Comp** and **Add** in the Function Selector. Select **Comp** by typing **C,** and then select the **Add** function by typing **A.** A small dialog box appears, requesting that the name of the desired gate be entered.

Type: `EXOR`

Press: ↵

Mouse The **Add** and **Comp** functions can be activated by moving the mouse pointer over each function and clicking. These functions are automatically activated, however, whenever you select a component from the Component Selector by clicking the component name. To add a component to a drawing, click the desired component in the Component Selector. Then either move the mouse pointer to the desired position in the Display area and click, or click and drag. To click and drag, drag the component into position by holding the left button down, and then release the button when the component is in place.

You can also move a component by using the mouse. This procedure is described in Tutorial 8.

An Exclusive Or gate has now been drawn. Notice that the gate is positioned with its top input located at the cursor position. The second EXOR gate can be added by positioning the cursor and pressing [Ins]. Move the cursor down about one grid by pressing [↓] once and then move the cursor about three-quarters of the way across the Display area by pressing [→] repeatedly (about 30 times).

Press: [Ins]

A second EXOR gate has now been added by using the [Ins] key. [Ins] will add whatever component was the last to be selected.

Now move the cursor into position in order to add the left AND gate. The cursor should be between half and three-quarters of the way down from the top of the Display area. To add this gate,

Type: `A`

Type: `AND2`

Press: [↵]

Now move the cursor to the right in order to add the second AND gate. To add this gate, simply press [Ins].

Finally, add the OR2 gate to the right of the second AND gate. First, type **A.** Then enter the gate name. The dialog box still contains the name AND2, so type **OR2** and then press [Del] to delete the “2” remaining from the AND2 gate. Then press ↵.

The drawing on your screen should now look like Figure 1.6. After practicing several other MICRO-LOGIC II functions on this drawing, we’ll save it for later use.

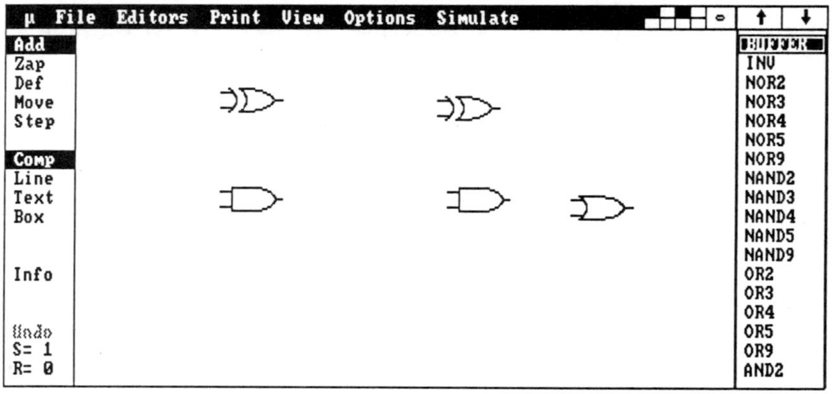

Figure 1.6

Deleting Components and Undoing Mistakes

To zap (delete) a specific component, select the **Comp** and **Zap** functions on the Function Selector and position the cursor on the component.

Position the cursor somewhere on the OR gate. Since Comp is already selected,

Type: `Z`

The OR gate should have disappeared. If it did not disappear, try repositioning the cursor (move it closer to the upper input of the gate); now type **Z** again.

If by accident you removed the wrong component, you can undo the last command executed by selecting the **Undo** function in the Function Selector.

Type: `U`

The OR reappears, since the Undo function undoes the last command to be performed, which in this case deleted the OR gate. The Undo function will undo other commands in addition to the Zap command.

> **Mouse** Click the **Zap** function in the Function Selector. Click the **Comp** function in the Function Selector. Move the mouse pointer to the component to be deleted, then click. If the wrong component is deleted, restore it by clicking **Undo**.

Another way to delete a component is to use the [Del] key. If you position the cursor over a component, select **Comp** in the Function Selector, and press [Del], the component will be deleted. Try this by positioning the cursor over the left EXOR gate and pressing [Del]. Then restore the gate by typing **U.**

Using the File Menu

The **File** menu contains commands that allow you to save and load drawings from the disk, as well as other file-related commands.

First make sure that your drawing looks like Figure 1.6. If it doesn't, correct the drawing as needed.

Updating Macros

As you progress in MICRO-LOGIC II, you'll use complicated components that are stored as macros. So that these macros will be ready when you need them, you should update them using option **7:Update macros** on the File menu. To open the File menu and activate this option,

Press: [Alt] F

Type: 7

A box will appear with the message **Updating macros,** also displaying the name of the file it is working on. This process will take a few minutes. You won't have to update the macros again unless you change one of them.

Defining the Default Disk Drive

Information can be stored or retrieved from any disk drive in the system. This is done by selecting option **6:Change data drive** on the File menu. Determine on what drive you wish to save your drawings. In a two-floppy system, this is normally drive B, and in a hard-disk system, drive C. To select this option,

Press: [Alt] F

Type: 6

In the middle of the display area a special dialog box called the **File dialog box** appears as shown in Figure 1.7. This box is used whenever a disk-related operation is performed. (Later, when you load an existing drawing from the disk, the File dialog box will contain a list of files.) At the top of the box is a letter that indicates the current data drive. If this drive is correct, press [Esc]. If you wish to change the active data drive, use the [↑] and [↓] keys to highlight the correct drive letter, then press [↵]. (The bubbles labeled **Change** and **Done** are used only by a mouse.)

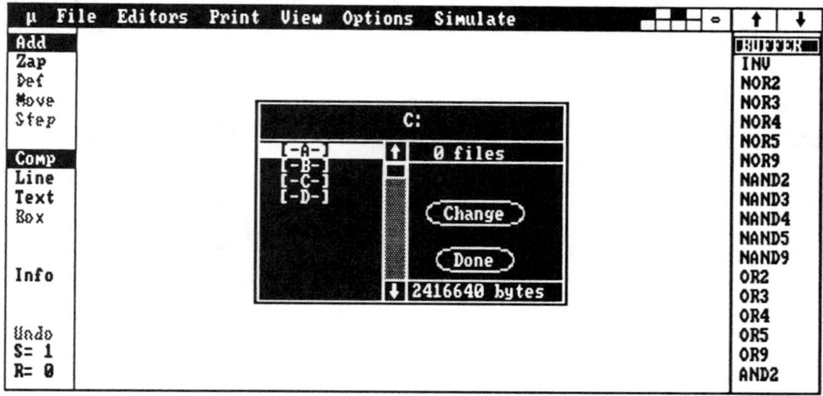

Figure 1.7

Mouse Refer to Figure 1.8 during the following discussion. You can scroll through the list of disk drive letters or files in three ways:

- Clicking the mouse over the up scroll arrow or the down scroll arrow will move the highlighting cursor up or down one position per click.
- Clicking and dragging the scroll box will move the highlighting cursor proportionately.
- Clicking the mouse in the scroll bar (outside the scroll box) will move the scroll box to the clicked position and move the highlighting cursor proportionately.

Once the desired disk drive or file is displayed in the File dialog box, it can be selected using any of the preceding three methods or by clicking the mouse directly on the desired file name. The command that activated the File dialog box can be completed by clicking the appropriate response bubble as in a regular dialog box.

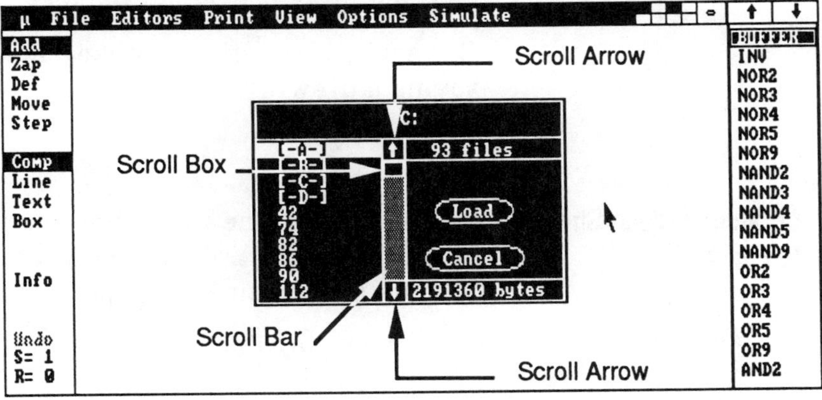

Figure 1.8

Note: If you have a system with a drive different from A, B, C, or D, you can still select that drive. Press [Alt] ***drive letter***, where ***drive letter*** is the desired drive letter. This will change the last drive letter in the File dialog box from D to your drive letter. Then select this drive as the default data drive.

Saving Drawings on Disk

You can now save your drawing on the default disk drive that you just defined. Use option **1:Save drawing** on the File menu. Give the drawing the name FULL_ADD:

Press: [Alt] F

Type: 1

Type: FULL_ADD

Press: [↵]

The drawing has now been saved on disk in a drawing file called FULL_ADD. MICRO-LOGIC II automatically adds an extension, DWG, to the file name. Thus the drawing is actually stored in a file with the filename FULL_ADD.DWG. When saving or loading a drawing, do not type the extension DWG.

Warning: When a drawing is saved, it will overwrite any existing drawing with the same name. You are not alerted that this will occur. Thus you must use caution when saving a drawing. Make sure that another drawing with the same name does not already exist on the disk.

It is a wise precaution to make a new copy of the data disk after saving a drawing.

Creating a Clean Sheet of Paper

When starting a new drawing, you must prepare a clean sheet of paper by using option **5:New drawing** in the File menu. To prepare a clean sheet of paper,

Press: [Alt] F

Type: 5

The Display area should now be clear. A new clean sheet of paper has been prepared and is ready to draw on. When you prepare a clean sheet of paper, any existing drawing in the Display area will be erased. If the current drawing was never saved or has been modified since it was last saved, you'll be warned of this before the drawing is erased.

To illustrate this warning, add any component anywhere in the Display area—this now is a drawing that has not been saved to disk. Attempt to create a clean sheet of paper by pressing Alt **F** and typing **5.** A dialog box appears that alerts you to the fact that the current drawing has changed since it was saved (see Figure 1.9). There are three possible responses to the question **Do you want to save it?**

- **Yes**, save the drawing on disk and then erase the drawing.
- **Cancel**, do not save the drawing and do not erase the drawing.
- **No**, do not save the drawing but do erase the drawing.

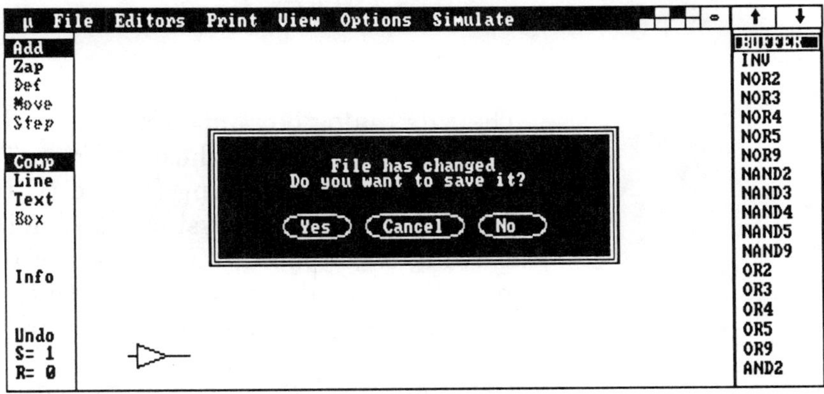

Figure 1.9

To respond, press Alt and the first letter of the desired response (**Y**, **C**, or **N**). In this case, since the existing drawing is something we don't want to save, press Alt **N.**

This special warning dialog box appears whenever both of the following are true:

- A drawing has never been saved or has changed since it was saved.
- A **load drawing** or **new drawing** command is executed or you attempt to exit MICRO-LOGIC II.

The warning message can be disabled by toggling option **8:File warning** in the File menu. Pull down the File menu by pressing Alt **F**. Note that option 8 has a check beside it. The check means that this feature is activated—that is, the warning dialog box is generated under the preceding conditions. If you now type **8**, the warning dialog box will not appear again. It is suggested, however, that you leave the warning dialog box activated. Therefore, simply press [Esc] to exit the File menu.

Loading an Existing Drawing from Disk

An existing drawing can be loaded from disk by using option **2:Load drawing** on the File menu. To illustrate this command, reload the drawing FULL_ADD. (This will also confirm that it was really saved.)

Press: [Alt] F

Type: 2

The File dialog box appears. It now displays all the drawing files that are stored on the data drive. The drawing to be loaded can be selected by using [↑] and [↓] to highlight the correct drawing and then pressing [↵]. (Another way to highlight a drawing is to type the first letter of its name. If the name of more than one drawing starts with that letter, type the letter again to highlight the second drawing.)

To terminate the load command without loading a drawing,

Press: [Esc]

Move the cursor to highlight the file FULL_ADD.

Press: [↵]

Erasing Drawings from Disk

If a particular drawing stored on disk is no longer needed, it can be erased by again using the File menu and selecting option **3:Erase drawing**. You will erase the file FULL_ADD, but first make sure that the current drawing in the Display area is a correct version of this file.

Press: [Alt] F

Type: 3

The File dialog box has now appeared. Select the file FULL_ADD by using the [↑] and [↓] keys to highlight it.

Press: [↵]

You will note that the file FULL_ADD no longer appears in the list of files. It has been erased from the disk.

Warning: You are given no warning that the selected drawing will be erased. If [↵] is accidentally pressed, whatever file was currently highlighted will be erased and cannot be recovered. Be very careful when using the Erase drawing command.

To exit the File dialog box,

Press: [Esc]

Since you just erased the FULL_ADD drawing, which you'll need later, save it again:

Press: [Alt] F

Type: 1

The filename FULL_ADD should be displayed as the name of the file to be saved. If it is not, type **FULL_ADD**. Press [↵] to complete the save.

Merging Two Drawings

Two existing drawings can be combined by using option **4:Merge drawing** in the File menu. This option merges a drawing from disk with a drawing in the Display area. The drawing from disk will be copied; the upper left corner of the drawing will be placed at the current cursor position. Let's demonstrate this by using our FULL_ADD drawing and merging the drawing SIMEX with it. Make sure that a copy of the drawing FULL_ADD is currently in the Display area. (Load the drawing from disk if necessary.) Use the cursor keys to move the cursor to a point about halfway between the bottom of the logic gates and the bottom of the Display area.

Press: [Alt] F

Type: `4`

A dialog box appears asking if you wish to **merge a circuit drawing file (Y/N)**. The default answer, **N**, is displayed. If you'd made a mistake and didn't want to merge drawings, you could press ↵ or type **N** to return to the Designer Display. In this case, however, we want to continue with the merge operation.

Type: `Y`

Another dialog box appears asking if you wish to **copy text from circuit file also (Y/N).** A drawing can be merged either with its text (labels) included or without them. The default answer is **Y**. We'll include the text.

Press: ↵

The File dialog box now appears. Using the cursor keys, select the file SIMEX and press ↵. You will notice that a new drawing has appeared under the FULL_ADD drawing. This is the SIMEX drawing, which has now been merged with FULL_ADD.

Since we won't need this merged drawing, throw it away by pressing Alt **F** and typing **5**. At the File Warning dialog box, press **N.**

Deactivating the File Dialog Box

The File dialog box can be deactivated during a load operation. Rather than selecting a file by using the File dialog box, you must type in the drawing name.

Pull the File menu down by pressing Alt **F**. Notice that option **9:File dialog box** is checked. This means that the File dialog box will appear when a file is loaded. Toggle this option by typing **9**.

To observe the result of this action, try loading the file FULL_ADD again. Press Alt **F** (notice that option 9 is no longer checked) and type **2**. You won't see the File dialog box, but rather a standard dialog box in which you must type the name of the desired drawing. Type **FULL_ADD** and press ↵.

If you prefer to use the File dialog box, pull down the File menu by pressing [Alt] **F** and toggle option 9 again by typing **9**.

Component Rotation and Reflection

In most logic diagrams, components are drawn with inputs on the left and outputs on the right. Sometimes it is necessary or useful to face components in a different direction. With MICRO-LOGIC II, components can be **rotated** into one of four positions: output facing right, left, up, or down. Components can also be **reflected** in the horizontal axis, thus giving a total of eight possible orientations.

At the bottom of the Function Selector is the current **rotation/reflection number** indicator, **R= x**, where x is a number between 0 and 7. This number indicates the rotation and reflection that will be given to a component when it's added.

To rotate or reflect a component, the current rotation/reflection number must be set to the desired value by using the number keys at the top of the keyboard. The eight possible values are listed in Table 1.2.

0 Output faces right, no reflection

1 Output faces down, no reflection

2 Output faces left, no reflection

3 Output faces up, no reflection

4 Output faces right, component is reflected

5 Output faces down, component is reflected (output ends up)

6 Output faces left, component is reflected

7 Output faces up, component is reflected (output ends down)

Table 1.2

For a symmetric device, such as an invertor, rotation/reflection values of 4–7 merely repeat orientations 0–3. Asymmetric devices such as flip-flops, however, can be oriented uniquely in all eight directions.

Start with a clean sheet of paper. If the Display area isn't clear,

Press: [Alt] F

Type: 5

Remember that if the current drawing has been changed since it was saved, or has never been saved, you will be asked what you want to do with it before the Display area can be cleared.

Position the cursor in the extreme upper left corner of the Display area using [↑] and [↓].

Type: C

Type: A

Type: JNC

Press: [↵]

This first component has been drawn with a rotation/reflection value of 0. Now add a second JNC flip-flop with a rotation/reflection value of 1.

Press: [→] 19 times

Type: 1 (don't use the numeric keypad)

Notice that R= 1 in the Function Selector, indicating that the current rotation/reflection value is 1.

Press: [Ins]

Now let's add six more JNC components to the drawing to illustrate the remaining six rotation/reflection orientations. (In the following steps you may position the cursor wherever you like. The suggested cursor positioning will allow you to add all eight components in sequence within the Display area without overlapping any two components.) Don't use the numeric keypad when typing the numbers 2 through 7.

Press: [→] 14 times

Press: ↓ 10 times

Type: 2

Press: Ins

Press: → 6 times

Type: 3

Press: Ins

Press: ↓ 20 times

Type: 4

Press: Ins

Press: ← 6 times

Type: 5

Press: Ins

Press: ← 11 times

Type: 6

Press: Ins

Press: ← 21 times

Type: 7

Press: Ins

The screen should look like Figure 1.10. Eight J-K flip-flops have been drawn in all eight combinations of rotation and reflection. Study these eight combinations and compare them with the definitions given in Table 1.2.

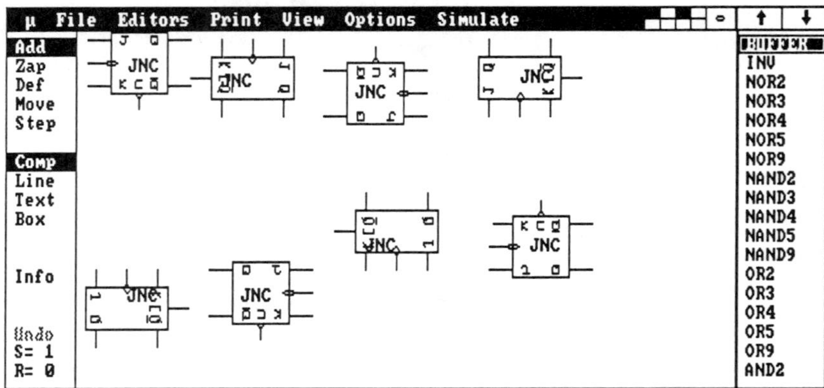

Figure 1.10

Mouse The current rotation/reflection number can be reset to zero by clicking the displayed value in the Function Selector. In addition, you can rotate and/or reflect a component when you add it using the click and drag method. While dragging the component, you can change the rotation/reflection number by clicking the right button. When you obtain the proper rotation/reflection and reach the proper position in the drawing, release the left button.

Option **5:Mouse Ratio** in the Options menu selects one of two ratios, high or low, for mouse movement. When this option is not checked, the mouse pointer will move twice as far per unit movement of the mouse as when the option is checked. The option toggles between the checked and unchecked setting every time it is selected from the menu.

Problems

1. Answer the following questions relating to the Component Selector.

   a. How many total components are available?
   b. How many Boolean gates are available?
   c. How many data channels are available?

d. How many clocks are available?
e. How many flip-flop macros are available?
f. How many integrated circuit packages are available?
g. Scroll through the Component Selector window or look in the Reference Section to identify whether the following components are contained in the library.
______ A 6-input NAND gate
______ A positive edge-triggered D flip-flop with preset and clear
______ A J-K master-slave flip-flop with preset
______ A 74139
______ A 74121

2. Draw an invertor in all eight possible rotations and reflections. Which rotation/reflection numbers result in equivalent drawings and why?

3. A positive edge-triggered D flip-flop with clear is shown in Figure 1.11 in one of the eight orientations. Draw this device somewhere on your screen with its orientation as shown in Figure 1.11.

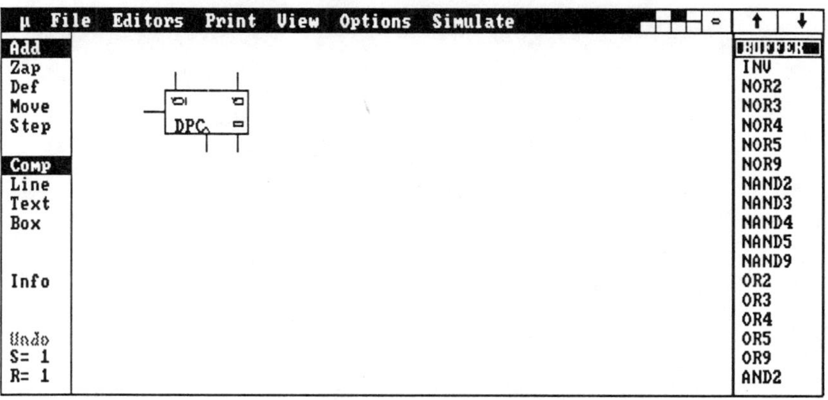

Figure 1.11

4. Clear the display area and draw an invertor somewhere on the screen using the Add command (do not use the Ins key). Now move the cursor to some other location in the display area; press ⇆ once. If you now press Ins, what component would you expect to be drawn? Try it and determine what component was actually drawn and why.

2 Using Lines and Text and Controlling the Display Area

2 Using Lines and Text and Controlling the Display Area

Tutorial 2 will teach you how to:

- Draw lines to connect component inputs and outputs
- Include text in order to label components, inputs, and outputs
- Control the positioning of the Display area
- Make hardcopies of your logic diagrams

Adding and Deleting Lines on a Logic Diagram

Components can be interconnected by drawing **lines** between the appropriate input(s) and output(s). A line represents the **wire** that connects two components in a real circuit. All points along the wire are assumed to have the same electrical potential and the same logical state at every instant in time. An interconnecting wire is also referred to as a **node**. The terms line, wire, and node are used interchangeably in this manual.

Adding Lines

You must keep several important points in mind when drawing lines:

- A line must begin and/or end on a component input or output or on another line.

- Lines that cross are not connected.
- Only horizontal and vertical lines are permitted; diagonal lines cannot be drawn.
- To begin a new line, type **A.** To change direction or to terminate a line, press [Ins]. The origin of a line is always marked with a small square.

Now let's interconnect the components in our FULL_ADD circuit. Load this drawing into the Display area by pressing [Alt] **F** and typing **2**. Then use the cursor keys to highlight the file FULL_ADD and press [↵].

Figure 2.1 shows the interconnections that must be made between the components. Let's first connect the output of the left EXOR gate to the top input of the right EXOR gate. First, select the **Line** function:

Type: `L`

Then move the cursor to the output of the left EXOR gate,

Type: `A`

A line has been started at the output of this gate. (You should notice a slight change in the cursor.) Now use the [→] key to move the cursor to the right. The line will become visible when you press [Ins].

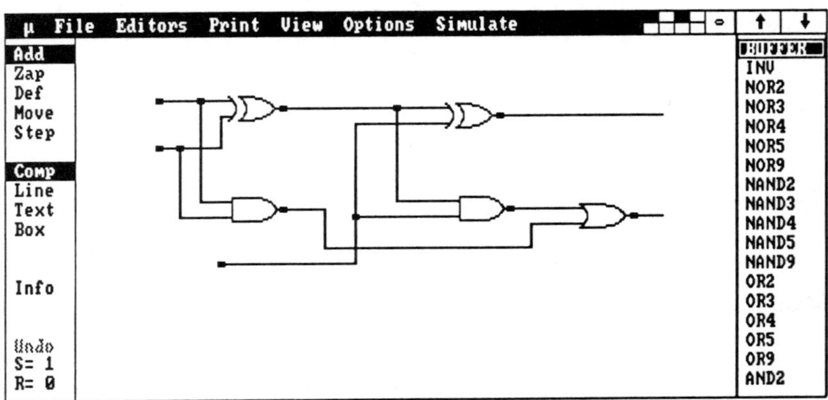

Figure 2.1

If you reach the input of the right gate, terminate the line by pressing [Ins]. This will happen only if the output of the left gate is exactly lined up with the input of the right gate. If they do not line up properly, use [→] to move the cursor partway across the screen toward the right gate.

Press [Ins] to change direction and then use either [↑] or [↓] to position the cursor directly opposite the input. Press [Ins] to change direction again, and use [→] to move the cursor to the input. Finally, press [Ins] to terminate the line.

Note: When adding lines, always type **A** to start a new line. If you press [Ins] instead, a line will be drawn from the endpoint of the last line drawn to the cursor.

Now let's connect this line to the top input of the right AND gate. Position the cursor directly on the newly inserted line about three-quarters of the way between the two EXOR gates. Type **A** to start the line and press [↓] until the cursor is directly opposite the top input to the right AND gate. Press [Ins] to change direction and then move the cursor right to the input by pressing [→]. Press [Ins] to terminate the line.

> **Mouse** To draw a line, select **Add** and **Line** in the Function Selector. Position the mouse pointer where the line should start and click the left button. Move the mouse as desired, either horizontally or vertically (but not both). To change direction or to terminate a line, click the right button.

You may have noticed that you cannot draw a diagonal line. For example, move the cursor to the output of the OR gate.

Type: `A` (to start a new line)

Press: [→] 3 times

Press: [↑] 4 times

Press: [Ins] (to terminate the line)

Note that although you attempted to draw a diagonal line, MICRO-LOGIC II adjusted the cursor to the left to form a vertical line.

Deleting Lines

Now let's try to delete the line you've just drawn. Position the cursor anywhere on the line, then type **Z** to Zap (delete) the line. (Instead of typing **Z,** you could have pressed [Del].) When the line-delete function is used, only a line segment will be deleted. That is, only the portion of the line marked by the cursor and bounded by a corner, endpoint, or starting point will be deleted. Deleting a line with several corners requires several deletes.

> **Mouse** To delete a line segment, select **Zap** and **Line** in the Function Selector. Position the mouse pointer on some part of the line segment that is to be deleted and click the left button.

Completing the Circuit

Now complete the wiring of the rest of the circuit. Table 2.1 lists suggested starting and ending points for each line and a suggested order for drawing the lines. It makes no difference in what order you draw the lines, and it makes no difference if you reverse the starting and ending points.

Leave some room between the unconnected lines on the inputs of the left EXOR and AND gates and the left side of the Display area. (Seven cursor grids is an acceptable distance.)

Start	**End**
Output of right EXOR	Right side of Display area
Output of right AND	Top input of OR
Output of OR	Right side of Display area
Unconnected line below left AND	Bottom input of right EXOR
Line just drawn	Bottom input of right AND
Output of left AND	Bottom input of OR
Unconnected line near left EXOR	Top input of left EXOR
Line just drawn	Top input of left AND

Start	End
Unconnected line near left EXOR	Bottom input of left EXOR
Line just drawn	Bottom input of left AND

Table 2.1

If you make a mistake when drawing a line, delete the incorrect line and add a correct one. **Undo** can be used to undo line commands as well as component commands. When undoing a line command, only the last line segment added or deleted will be affected. If the cursor cannot be correctly positioned on a line or component, check the cursor grid control in the Function Selector (S = 1). If S is not equal to 1, read the section on Cursor Grid Control in this chapter in order to set S to 1.

Labeling Drawings

Text or **labels** can be added to a logic diagram. When you label the devices and signals in the circuit, it is easier to distinguish the various components and signals. Also, when you use MICRO-LOGIC II to simulate a circuit, you can select appropriately labeled signals to be monitored. (It may not be desirable or possible to monitor all circuit signals during a simulation.) Throughout this discussion the terms text and label will be used interchangeably.

Adding Text

Now let's add the title FULL ADDER CIRCUIT to our drawing. To add text, place the cursor where you want the text to start. The text will be added just above the cursor and will extend to the right. Now select the **Text** and **Add** functions in the Function Selector. (If **Add** is already selected, you must select it again.)

A dialog box will now ask you for the text to be added. Whether typed in uppercase or lowercase, text will always appear in uppercase on the drawing; any leading spaces will be deleted.

Add the text FULL ADDER CIRCUIT by moving the cursor to the left edge of the Display area just below the circuit.

Type: `T`

Type: `A`

Type: `FULL ADDER CIRCUIT`

Press: ⏎

The Display area should now look like Figure 2.2.

Instead of typing **T** and then **A** to insert text, you could have typed **T** and then pressed Ins. (If **Text** is already highlighted, you need not select it again.)

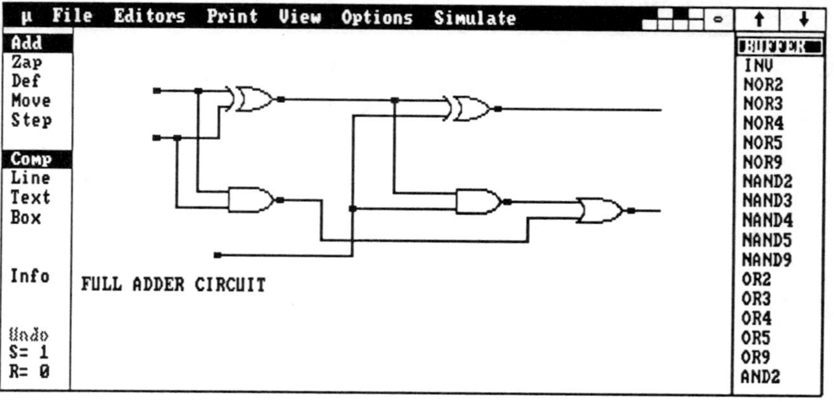

Figure 2.2

Mouse To add text, select **Add** and **Text** in the Function Selector. Position the mouse pointer where the text is to be placed and click the left button. Type the desired text in the dialog box and either press ⏎ or click the mouse anywhere in the Display area.

With a little practice, you can place text wherever it is needed. If the text is not positioned properly, you can delete it and add it again in the correct position. To delete the label FULL ADDER CIRCUIT, position the cursor somewhere over the text and use the **Zap** function.

Type: `Z`

You can also delete text by pressing [Del] instead of typing **Z**. In addition, you can use the **Undo** function to undo either text additions or text deletions.

Try positioning the label so that it is centered under your drawing by using the Add and Zap functions.

Mouse To delete text, select **Zap** and **Text** in the Function Selector. Position the mouse pointer over the text to be deleted and click the left button.

Labeling Lines for Simulation

A signal that will be monitored during simulation should be labeled on the line so that the computer will associate the label with the line. To associate a label with a line, position the cursor exactly on the line when you insert the text. Remember that the text will be inserted slightly above and to the right of the cursor.

Now let's add the labels SUM, CARRY, CIN, IN1, and IN2 as shown in Figure 2.3.

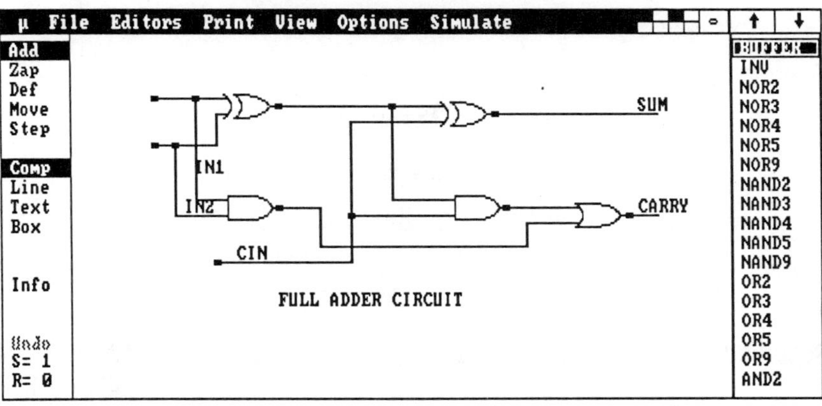

Figure 2.3

Move the cursor near the right end of the line connected to the right EXOR. Make sure that the cursor is actually on the line. Since **Text** is already selected,

Press: Ins

Type: `SUM`

Press: ↵

Notice that the label SUM now appears over the line. Add the labels CARRY, CIN, and IN2 in a similar fashion, making sure that the cursor is positioned exactly over the line each time.

The four labels just added were added to horizontal lines. Labels can also be added to vertical lines. Move the cursor so that it is on the vertical line connected to the top input of the left AND gate. To add the text IN1, move the cursor onto the line, then

Press: Ins

Type: `IN1`

Press: ↵

The label of a vertical line is functionally no different from the label of a horizontal line.

Your drawing should now look like Figure 2.3. If it doesn't, correct it so that it does. We'll save this drawing for future use.

Press: Alt `F`

Type: `1`

Type: `FULL_ADD`

Press: ↵

Reserved Text and Ties

You may not use the label **PIN** or **PIN*n***, where ***n*** is a number, in a drawing. This special label can be used only when defining a macro (see Tutorial 7).

In complex drawings, sometimes a continuous line cannot be drawn between all connecting points. The **tie** interconnect feature allows line segments that are not physically connected in the drawing to be logically and electrically connected. Any label associated with a line that begins with the reserved character "/" indicates a tie connection. Any line segments with the same label beginning with this character are logically and electrically connected. For instance, in Figure 2.4, the two line segments labeled **/ABC** are logically and electrically connected even though they're not physically connected.

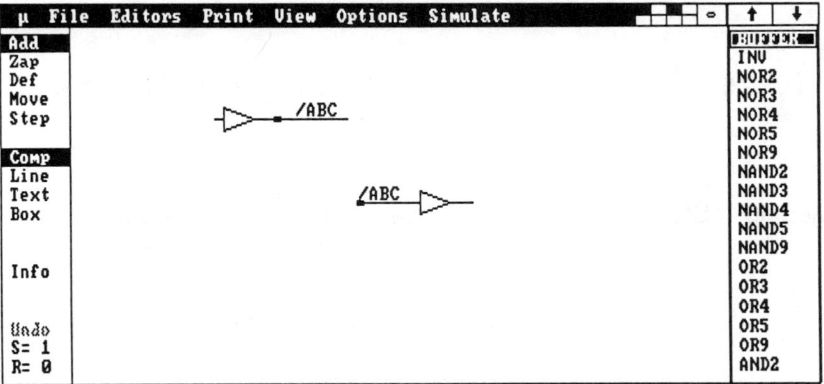

Figure 2.4

Moving the Drawing Paper

The Display area shows only one small part of the total drafting paper available for drawing. The Display area represents what is known as one **page** of the paper. A group of pages forms a **quadrant**, and four quadrants form the entire drafting surface. (The number of pages in a quadrant depends on your graphics adapter.)

To find out which page and quadrant are currently being displayed, examine the **Drawing View controls** as shown in Figure 2.5. Figure 2.5 shows a system with four pages per quadrant. The upper left page of the upper left quadrant is currently displayed. Other pages and quadrants can be accessed with the scroll feature either off or on.

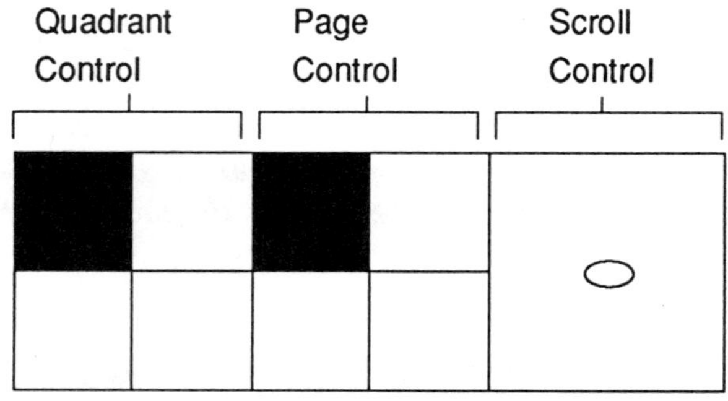

Figure 2.5 Drawing View Controls

Method 1: Scroll Off

With the scroll off, use the four cursor keys—[↑], [↓], [←], and [→]—as well as [PgUp], [PgDn], [Home], and [End]. (The scroll is toggled on and off by pressing [Scroll Lock]. If you haven't pressed [Scroll Lock] since starting MICRO-LOGIC II, then the scroll is off.)

The cursor keys will move the cursor across the page. When the cursor reaches the edge of the Display area, the paper will be scrolled. [PgDn] will move the Display area down one page. [PgUp] will move the Display area up one page. [End] and [Home] will move the Display area right one page and left one page, respectively.

The currently displayed page should be the upper left page of the upper left quadrant. If the drawing FULL_ADD isn't displayed, load it. With the scroll off and while observing the Drawing View controls,

Press: [PgDn]

The logic diagram has disappeared and the Drawing View controls indicate that the paper is positioned in the lower left page. (If the page didn't change, the scroll is on. Press [Scroll Lock] and repeat the previous step.)

Press: PgUp

You should have returned to the upper left page and should see your drawing again. To move to the upper right page, press End. To move back to the upper left page, press Home.

You can test the functioning of the PgUp, PgDn, Home, and End keys as much as you like. Just make sure when you're done that you've returned to the drawing in the upper left page by using these keys and/or the cursor keys.

Notice that although different pages were brought onto the screen, you remained in the original quadrant. Using the PgUp, PgDn, End, and Home keys with the scroll off will not move the paper out of the current quadrant.

Mouse The Drawing View controls, shown in Figure 2.5, can also be used to select the page and quadrant to be viewed. To select a quadrant, position the mouse pointer in the appropriate quadrant box in the Drawing View controls and click. To select a page, position the mouse pointer in the appropriate page box in the Drawing View controls and click.

The large box in the Drawing View controls with a circle in its middle is the **scroll control**. Clicking in the upper right corner of the box will scroll the drawing to the right and up. Likewise, clicking in the lower left of the box will scroll the drawing to the left and down. The direction of the click from the center circle determines the direction of the scroll. The distance of the click from the center circle determines the scroll size and hence the speed.

Method 2: Scroll On

With the scroll on, use the cursor keys—↑, ↓, ←, and →—as well as PgUp, PgDn, Home, and End.

The cursor keys no longer move the cursor. Instead they scroll the paper, regardless of the position of the cursor. In addition, PgUp, PgDn, End, and Home don't move the paper in page steps. Instead these keys scroll the paper diagonally up and right, down and right, down and left, or up and left, respectively. Perform the following sequence of steps and observe what happens in the Display area.

Press: [Scroll Lock]

Press: [PgDn] 3 times

Press: [PgUp] 3 times

Press: [End] 3 times

Press: [Home] 3 times

Press: [Scroll Lock]

Mouse The paper will scroll if the mouse pointer is placed at the edge of the Display area while the scroll is on. If the mouse pointer is placed in a corner of the Display area, the paper will scroll diagonally. Scrolling is activated by the mouse in only a small region around the edge of the Display area. If the page doesn't scroll, try pressing [Scroll Lock] and/or slightly repositioning the mouse pointer.

Cursor Grid Control

To move the cursor all the way across the drawing paper takes a noticeable time: The cursor moves only one grid with each press of a cursor key. This time can be reduced by changing the number of grids that the cursor moves per keypress. The **S=** near the bottom of the Function Selector indicates the current **grid step number**. (S should currently be equal to 1.) The number of grids moved can be changed by pressing [+] and [−] on the numeric keypad. Pressing [+] doubles the value of S; pressing [−] resets S to 1. Each doubling of S results in the cursor moving twice as many grids per keypress.

Press: [+] 2 times

Notice that S equals 4 because it was doubled twice. Now try moving the cursor using any of the cursor keys. You should observe that the cursor now moves four times as far per keypress as before. Press [−] and try moving the cursor. S has been reset to 1, so the cursor now moves only one grid per keypress.

S can reach a maximum value of 64. If 64 is doubled, the value is reset to 1. You'll probably want to set the grid step number to 1 for maximum resolution while you're drawing. When moving around the paper, however, increase the grid step number as appropriate to facilitate fast movement.

Mouse To reset the grid step number to 1, position the mouse pointer over the grid step number indicator in the Function Selector, **S=**, and click.

Printing Drawings

A drawing in the current quadrant shown in the Display area can be **printed**. Only the currently displayed quadrant can be printed. Make sure that your printer is one specified in the Getting Started section and that it is properly connected as device lpt1. To print a drawing, select option **1:print drawing** from the **Print** menu.

Press: [Alt] `P`

Type: `1`

A printer options dialog box has appeared. Drawings can be printed in either a small or a large version. A small drawing will contain the entire quadrant on one page. If you choose the large version, approximately the top two-thirds of the quadrant will be drawn on one page; the bottom third will be drawn on another page. If you choose a large drawing, you can also choose whether the printing will be dark or light. Darker printing is denser and generally takes longer to print than light printing.

To choose a printer option, type the number indicated on the menu. To exit from this menu without printing, press either [Alt] **C** or [Esc]. Printing a drawing takes some time. If after the printer begins to print, you decide that you don't want a hardcopy, you can cancel the printing by pressing [Esc].

Quadrant and Full Drawing Views

Normally only one page of a drawing is visible on the screen. It is possible to view the entire quadrant or drawing if desired, however, by selecting the appropriate option in the **View** menu. The quadrant view is option **2:Quadrant** in the View menu.

Press: [Alt] V

Type: 2

The entire quadrant is displayed in the upper left corner of the Display area. To exit this view, press [↵]. The entire drawing can be viewed by selecting option **3:Quadrants.**

Press: [Alt] V

Type: 3

Now the entire drawing appears in the upper left corner of the Display area. To exit this view, press [↵].

Problems

1. Does the rotation/reflection number affect the positioning of text?
2. Explain the difference between moving the cursor keys with the scroll on and with the scroll off.
3. Answer the following questions about drawing lines. Assume that **Line** has been selected on the Function Selector and that **A** has been typed just prior to performing each of the actions described.
   a. If [→] is pressed four times and [↑] is pressed twice, in what direction will a line be drawn when [Ins] is pressed?
   b. If [↑] is pressed five times and [←] is pressed twice, in what direction will a line be drawn when [Ins] is pressed?

c. If [→] is pressed three times and [↓] is pressed three times, in what direction will a line be drawn when [Ins] is pressed? Does the order in which [→] and [↓] are pressed make any difference?

4. Draw logic diagrams for the following Boolean expressions. Do not simplify or change the expressions.

   a. OUT1 = AB ⊕ C

   b. OUT2 = (ABC)'+(B'C')

   c. OUT3 = A'B + AB'

3 Circuit Inputs and Reports

3 Circuit Inputs and Reports

In Tutorial 3 you'll learn how to:

- Specify signal inputs (data channels) for a circuit
- Define the state of the data channels
- Generate reports that describe the logic circuit

Circuit Inputs

In the first two tutorials we used MICRO-LOGIC II only to draw logic diagrams. Although a logic simulator such as MICRO-LOGIC II can be used for this purpose alone, its real purpose is to perform logic simulation.

The first step in simulating the functioning of a circuit is to define the circuit by drawing it. The next step is to define the inputs of the circuit. Then the circuit can be simulated and the generated outputs can be compared with the defined inputs.

In MICRO-LOGIC II, a data input is called a **data channel.** With the software as purchased, as many as 16 different data channels can be defined for each circuit. An additional 16 data channel components can be created using the Component editor described in Tutorial 7 (for a total of 32 possible data channels). Data channels have component names of DATA1,

DATA2, ..., DATA16.

In order to define a data input, two conditions must be met:

- The data channel must be included in the logic drawing.
- The state of the data channel signal must be defined.

Including Data Channels in Logic Drawings

Let's add data channels to our drawing FULL_ADD. Load the drawing by pressing [Alt] **F** and typing **2.** In the File dialog box highlight the file FULL_ADD by using [↑] and [↓] ; then press [↵] .

Figure 3.1 shows what the circuit will look like when we're done. The circuit has three inputs—IN1, IN2, and CIN—so we must add three data channels. Notice that the data channel graphic symbol is a box enclosing a small waveform, with the text **DATA*n*** above it, where ***n*** equals the data channel number.

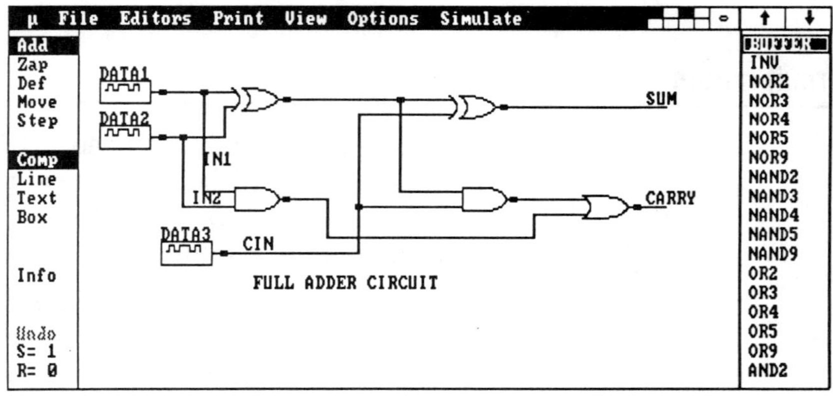

Figure 3.1

Move the cursor to the left end of the line where DATA1 is to be added. This is the horizontal line that is connected to the top input of the left EXOR gate. To correctly position the cursor in order to add the data channel,

Press: [←] 6 times

Press: [↑]

Note: If you don't have enough room at the left margin to fit the data channel, try adding the data channel with a rotation/reflection value of 1 or 3. If there still isn't enough room, you may have to redraw part of your circuit.

The data channel is added to the drawing at the cursor position just like any other component.

Type: `C`

Type: `A`

Type: `DATA1`

Press: ↵

The component DATA1 has now been added to the drawing. This represents a specific input signal (not yet defined) that will be applied to the signal line IN1 when a simulation is performed. Since a data channel is a component, all functions that can be performed on a component can be performed on a data channel, including deleting, rotating, and reflecting.

Note: For keyboard users, a new list of components is now displayed in the Component Selector. As mentioned in Tutorial 1, however, this part of the component list is not available for viewing by pressing ⇆. It will appear only when one of the components listed on the screen is added to a drawing (or when a mouse is used).

Now add the other two data channels as shown in Figure 3.1. Remember that they can be positioned accurately if you place the cursor at the end of the line and then move the cursor left six grids and up one.

When finished, save the drawing:

Press: Alt `F`

Type: `1`

Type: `FULL_ADD`

Press: ↵

Figure 3.2 illustrates the basic format of a data channel. A data channel is a binary signal that changes state only after a predefined time called a **clock cycle**, where the length of the clock cycle is defined during simulation. A data channel can be defined to be from 1 to 1024 clock cycles in length. The **Pattern editor** is used to define data channels. The Pattern editor is called from the **Editors** menu by selecting option **3:Patterns**.

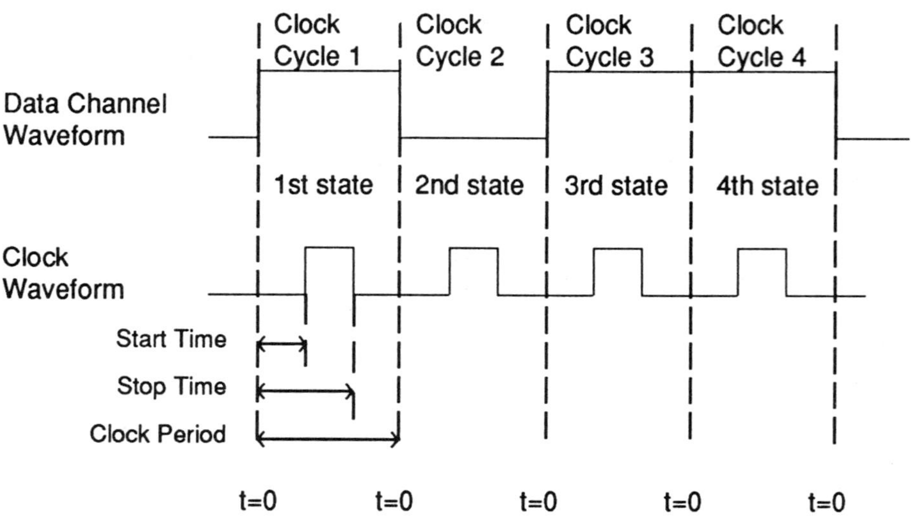

Figure 3.2

Press: Alt E

Type: 3

Your screen should look like Figure 3.3, which shows the Pattern editor screen. All 16 data channels can be defined. These are labeled as DATA 1 through DATA 16 down the left side of the screen. (Data channels 17 through 32 have not yet been created by using the Component editor.) Each pattern can be as many as 1024 clock cycles long.

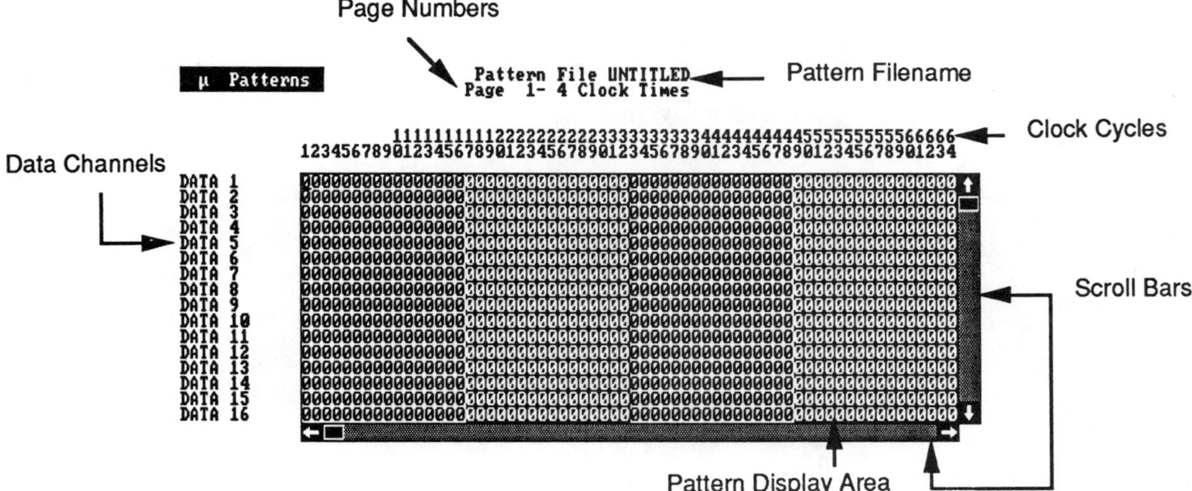

Figure 3.3

You can define a data channel either by direct keyboard editing or by using options found in the Patterns menu. For keyboard editing, you can select a particular data channel by using the ↑ and ↓ keys to position the cursor. Initially the cursor will be in the upper left corner of the **Pattern Display area**. Try pressing ↓ and ↑ to move the cursor to different data channels.

Even though a pattern is 1024 clock cycles long, only 64 clock cycles can be viewed at any one time. The currently viewable clock cycles are identified across the top of the Pattern Display area by a series of vertical numbers, with the most significant digit of the number on top. Three examples of this numbering are shown in Table 3.1.

3 = Clock Cycle 3

4

7 = Clock Cycle 47

1

0

2

1 = Clock Cycle 1021

Table 3.1

A group of 16 clock cycles is defined to be a **page**. Figure 3.3 displays pages 1 through 4 (or clock cycles 1 through 64) of data channels 1 through 16. The current pages being viewed are indicated at the top of the screen as **Page 1- 4 Clock Times.**

The remaining clock cycles can be viewed by using the → and ← keys, with scroll either on or off. Try moving the cursor horizontally by pressing → and ← with scroll on and with scroll off. Notice that the page numbers and clock cycle numbers change.

Mouse The mouse can be used to scroll the Pattern Display area. The horizontal scroll bar (at the bottom of the Pattern Display area) scrolls the pattern area left and right. The vertical scroll bar (at the right edge of the Pattern Display area) scrolls the pattern area up and down. The scroll bars work the same way as the scroll bar in the File dialog box.

Direct Keyboard Editing

The most direct way to define the states of a data channel is to type in the value of the data channel for each clock cycle. This is done by moving the cursor to the data channel and clock cycle to be defined and typing the binary sequence.

Using [↑] and [←], move the cursor to DATA 1, clock cycle 1.

Type: `101111000101010101`

Notice that this pattern has been entered into Data Channel 1, Clock Cycles 1 through 18. This is a practical method for defining data channels when only a few clock cycles must be defined.

> **Mouse** The mouse can be used to move the cursor within the Pattern Display area by positioning the mouse pointer at the desired spot and then clicking. Entries can then be edited with the keyboard by typing 1s and 0s as desired.

The Patterns Menu

Several options are available for defining patterns using the **Patterns** menu. Let's define several data channels using this menu. To pull down the menu,

Press: [Alt] P

Four of the options on this menu—**R:Random pattern**, **B:Binary pattern**, **I:Invert pattern**, and **S:Set a block**—permit the rapid generation of some useful input patterns.

Random Pattern

Option **R:Random pattern** generates a random string of 1s and 0s in all 1024 clock cycles of any data channel. We will now define Data Channels 1 and 2 to be random signals.

Type: R (to select the random pattern option)

Type: 1 (to choose Data Channel 1)

Press: ↵

Type: P (to pull down the Patterns menu)

Type: R (to select the random pattern option)

Type: 2 (to choose Data Channel 2)

Press: ↵

The patterns for Data Channel 1 and Data Channel 2 now consist of random binary sequences. If no particular pattern is needed, this is a quick way to generate a complete pattern of 1024 clock cycles.

Binary Pattern

You can generate binary patterns by using option **B:Binary pattern**. A binary pattern consists of alternating strings of 1s and 0s. The number of 1s and 0s will depend on a **binary divider** value that must be chosen. The binary divider is a power of 2 between 2 and 1024. A value of 2 will generate the sequence 010101...; a value of 4, the sequence 00110011...; a value of 8, the sequence 0000111100001111...; and so on. A binary pattern always begins with a 0.

Let's define Data Channels 3, 4, and 5 to be binary patterns with divider values of 2, 4, and 8, respectively. To define Data Channel 3 with a divider value of 2,

Press: Alt P (to pull down the Patterns menu)

Type: B (to select the binary pattern option)

Type: 3 (to choose Data Channel 3)

Press: ↵

Type: 2 (to set the divider to 2)

Press: ↵

Repeat these steps, modified as necessary, to define Data

Channel 4 with a divider of 4 and Data Channel 5 with a divider of 8. When you're finished, the screen should look like Figure 3.4 (although Data Channels 1 and 2 may be different since they are random patterns).

Figure 3.4

Note: If you make a mistake when selecting patterns from the menu, you can cancel the selection by pressing [Esc]. When typing in the channel number or the binary divider, you can use the cursor keys—[↑], [↓], [←], and [→]—as well as [Del], [Ins], and [←] to edit the entry.

Invert Pattern

Option **I:Invert pattern** will invert an existing pattern in any data channel. We will define Data Channel 6 to be a binary pattern with a divider of 8 that begins with 1s, by defining a binary pattern and then inverting it. Define the binary pattern:

Press: [Alt] `P`

Type: `B`

Type: `6`

Press: [↵]

Type: `8`

Press: [↵]

Data Channel 6 should now be the same as Data Channel 5. Now we can invert it:

Press: [Alt] `P`

Type: `I`

Type: `6`

Press: [↵]

Notice that Data Channel 6 has been inverted from its original value. (Verify by comparing with Data Channel 5.)

Mouse A single clock cycle or a row of clock cycles can be inverted with the mouse. Position the mouse on the appropriate clock cycle. To invert just one clock cycle, click the right-hand button. To invert a row of clock cycles, click the right-hand button and drag the mouse across the desired string.

Set a Block

The final method of defining data channel patterns is to use option **S:Set a block**, which defines a block of data channels/clock cycles to be all 0s or all 1s. First pull down the Patterns menu by pressing [Alt] **P**; then select the **Set a block** option by typing **S**.

The Set dialog box shown in Figure 3.5 appears. The cursor is in the small vertical box at the right side of the dialog box. The four entries in this box can be edited by using the four cursor keys and the [←], [Del], and [Ins] keys. For example, let's set Data Channels 1 through 3, pages 1 through 1 (that is, only page 1) to 0. We can do this by defining the first source as 1, the last source as 3, the first page as 1, and the last page as 1. With the cursor under the 1 in the box at the end of the "first source" line,

Type: `1`

Press: [↵]

Type: 3

Press: [Del]

Press: [↵]

Type: 1

Press: [↵]

Type: 1

Press: [Del]

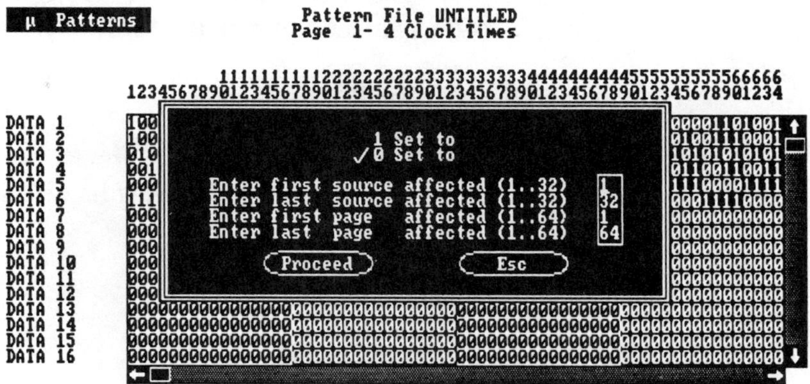

Figure 3.5

The **Set a block** command is activated as follows:

Press: [Alt] P

Note: If you make an error and want to quit without changing any patterns, you can press [Esc] instead of [Alt] **P**.

Notice that Data Channels 1, 2, and 3, page 1 (Clock Cycles 1 through 16) are set to zero (see Figure 3.6).

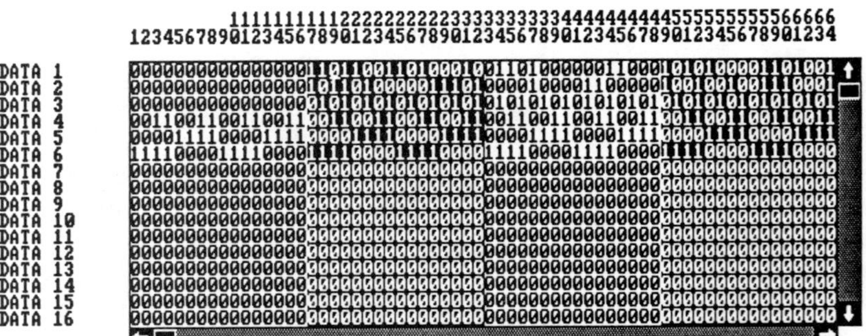

Figure 3.6

Now let's set Data Channels 4 through 7, pages 3 and 4, to logic 1. Again select the **Set a block** option:

Press: [Alt] P

Type: S

The Set dialog box should now be on the screen and can be set as follows:

Type: 4

Press: [↵]

Type: 7

Press: [↵]

Type: 3

Press: [↵]

Type: 4

Now only option **1 Set to** remains to be specified:

Press: [Alt] 1

(**0 Set to** can be selected by pressing [Alt] **0**.)

Notice that **1 Set to** is now checked. Press [Alt] **P** to complete the operation. Notice that Data Channels 4 through 7, pages 3 and 4 (Clock Cycles 33 through 64), have been set to 1.

Saving and Loading Pattern Files

Patterns to be used to define data channels must be saved to disk. This is done by selecting option **1:Save pattern** in the Patterns menu. Let's save this pattern under the filename TRIALPAT.

Press: [Alt] P

Type: 1

Type: TRIALPAT

Press: [↵]

The displayed pattern has now been saved under the name TRIALPAT. MICRO-LOGIC II automatically attaches the extension .PAT to the filename; the file is actually stored as TRIALPAT.PAT in a manner similar to files you have drawn. Notice that at the top of the screen the Pattern file is now labeled as TRIALPAT rather than UNTITLED as before.

Warning: If a Pattern file with the same name already exists on the disk, a Save operation will overwrite the disk file; no warning will be given.

A Pattern file, once saved, can be loaded from memory for re-editing or inspection by using option **2:Load pattern**:

Press: [Alt] P

Type: 2

When the File dialog box appears, use [↑] and [↓] to highlight the pattern file BINARY, and then press [↵]. BINARY is a predefined pattern that comes with MICRO-LOGIC II. Figure 3.7 shows this file, which is a series of binary patterns with dividers of 2, 4, 8, 16, 32, 64, 128, 256, 512, and 1024. We'll use Data Channels 1, 2, and 3 of this pattern when we simulate our full adder circuit in Tutorial 4. Notice that Data Channel 3 defines CIN, Data Channel 2 defines IN2, and Data Channel 1 defines IN1. In addition, these patterns match our previous assignments for CIN, IN2, and IN1 (see Table 1.1).

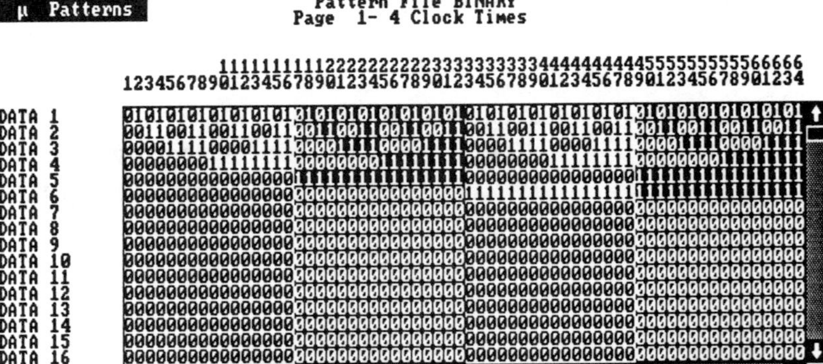

Figure 3.7

Just as when drawing files are loaded, when a pattern is loaded from disk it overwrites the current pattern. If the current pattern has not been saved or has been modified since saving, a dialog box will give you a chance to save the file before continuing.

Printing and Erasing Pattern Files

If desired, a hardcopy of the pattern can be made by selecting option **H:Hardcopy**.

Press: [Alt] P

Type: H

All 1024 clock cycles of all 32 data channels will be printed. This takes some time and paper to complete. The printing can be halted at any time by pressing [Esc].

You can erase a previously stored pattern from the disk by using option **3:Erase pattern**. Since the pattern file that we created and stored in this tutorial will not be needed later, let's erase it from the disk. (If you want to keep the pattern file, skip the next sequence of commands.)

Press: [Alt] P

Type: 3

The File dialog box now appears. To erase a previously stored pattern, simply highlight the correct pattern and press [↵]. To exit from the File dialog box without erasing a file, press [Esc].

Highlight the file TRIALPAT and press [↵]. If the menu does not close, press [Esc].

Warning: As when you erase a drawing file, you are given no warning that a file will be erased after the File dialog box appears. Once [↵] is pressed, the highlighted pattern will be erased and cannot be recovered.

To return to the Designer Display screen, press [Esc].

Longer Data Channel Patterns

Data channel patterns are limited in duration to 1024 clock cycles. During simulation it may be desirable to use longer patterns: This is possible but only by defining a set of pattern files.

For example, say you want to simulate a circuit that is to last for 3072 clock cycles (3 x 1024). Assume that the pattern file is to be called CKT. The pattern file CKT alone will only define 1024 clock cycles. If a second and third pattern file are defined with names CKT1 and CKT2, however, the simulation can be extended. When the simulation is performed, data is first taken from the pattern file CKT. Then file CKT1 is used, and then CKT2. This could continue up to a maximum value of 31, that is, through pattern file CKT31.

Displaying Labels

If you're still in the Pattern editor, exit by pressing [Esc]. The Designer Display should now be on your screen with the circuit FULL_ADD displayed. If FULL_ADD isn't displayed, load it from the disk.

Two types of text are currently displayed on the screen. One type consists of labels that you added to the drawing (for example, IN1 and SUM); the second type, of labels that are added by MICRO-LOGIC II when a component is added (for example, DATA1 and DATA2). Both types of labels can be temporarily removed from the drawing if desired. These labels are either displayed or hidden by toggling options in the **View** menu. To temporarily remove the labels you drew, toggle option **5:Show grid text**. First pull down the View menu:

Press: [Alt] V

Notice that menu options 4 and 5 are checked and thus are currently active.

Type: 5

Typing **5** deactivates this option, which means that your labels won't be displayed on the screen. To remove the component labels, select option **4:Show comp text**.

Press: [Alt] V

Type: 4

Option 4 has been deactivated, and component labels DATA1, DATA2, and DATA3 have disappeared. Your circuit should now look like Figure 3.8.

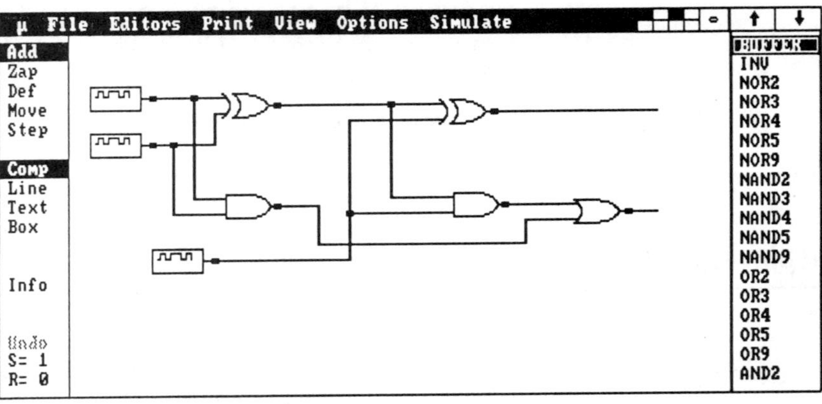

Figure 3.8

A third type of text is possible using MICRO-LOGIC II; it can be displayed by selecting option **6:Show comp nos**. Component numbers are assigned to each component and its corresponding output node by MICRO-LOGIC II. Although MICRO-LOGIC II needs these numbers to perform a simulation, they're not usually included in a standard logic drawing. (It is sometimes helpful to know these numbers when certain other MICRO-LOGIC II features are used.) The numbers are assigned when a component is added to a drawing, so the exact numbering of your components will depend on the sequence in which you added each component. To show the component numbers,

Press: [Alt] V

Type: 6

To restore your labels and the component text, simply enter the View menu twice, typing **4** once and **5** once. You can also remove the component numbers by pulling down the View menu and typing **6**.

Reports

MICRO-LOGIC II can generate a variety of reports that can be helpful in designing and debugging circuits. The reports are generated via the **Print** menu option **2:Print reports**.

It is suggested that during the next few steps you remove your labels and the component text from the drawing and that you display the component numbers, as explained in the preceding section. You'll find it helpful to see the component numbers. (If one of your labels or component text is positioned in the same space as a component number, the component number won't appear.)

After removing your labels and the component text and displaying the component numbers, select option 2 of the Print menu:

Press: [Alt] P

Type: 2

The **Report** menu now appears as shown in Figure 3.9. Selections are made from the Report menu by pressing the first letter of the desired activity. For example, the report can be sent to **Disk** by typing **D**, to the **Screen** by typing **S**, or to the **Printer** by typing **P**. A selected item is indicated on the menu by a check mark. Three types of reports can be generated: a Netlist report, an Unconnected Pins report, and a Capacity report.

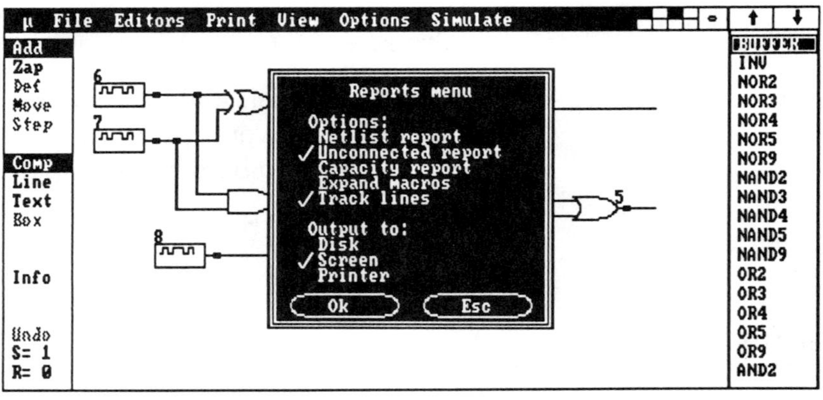

Figure 3.9

Netlist Report

Select the **Netlist report** from the Report menu by typing **N**. A Netlist report lists all components used in the drawing and their network connections (netlist). Be sure that **Screen** is checked. To generate the report,

Press: [Alt] o (for Ok)

While the report is being generated, for a very short time you'll see your logic diagram, upon which the nodes will be traced with small cursor squares. The node tracing occurs because the **Track lines** option is selected in the Report menu. This helps you visually identify lines that aren't connected.

Figure 3.10 shows the Netlist report for this circuit. At the top of the report is a list of the components in the diagram. The number on the left indicates the component number. To the right of the component number is the component type (such as EXOR or DATA1). To the right of the component type is a list of node numbers. The node numbers indicate all the other components to which the inputs to this component are attached—except for the last number, which indicates the output of the component. Remember that the numbering of your components may differ from those of Figure 3.10 if you added your components in a different order.

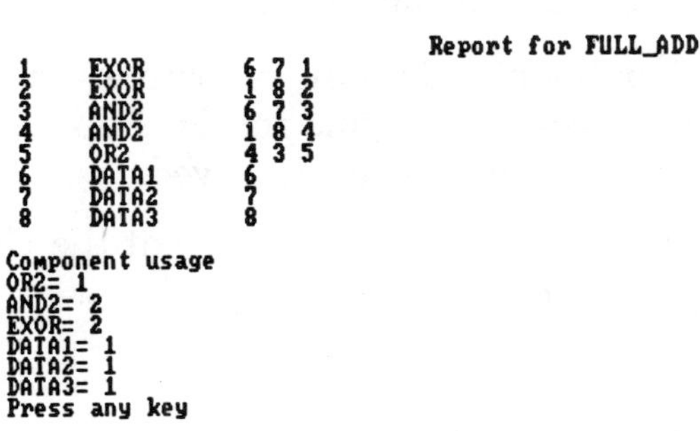

```
                                   Report for FULL_ADD
1   EXOR    6 7 1
2   EXOR    1 8 2
3   AND2    6 7 3
4   AND2    1 8 4
5   OR2     4 3 5
6   DATA1   6
7   DATA2   7
8   DATA3   8
Component usage
OR2= 1
AND2= 2
EXOR= 2
DATA1= 1
DATA2= 1
DATA3= 1
Press any key
```

Figure 3.10

If one of the node numbers of a device is 0, one of the inputs of the component is not connected. This usually indicates a wiring error.

At the bottom of the Netlist report is a report on component usage, indicating the number and type of devices used. You're limited to a certain number of components, lines, text, and so on. These limits are listed in the Reference Section.

After you finish examining the Netlist report, you can return to the Designer Display by pressing any key.

Unconnected Pins Report

The **Unconnected Pins report** identifies component pins that aren't connected. You'll find it helpful to generate this report after adding wires to a diagram to make sure that the wires were properly added.

Press: [Alt] P

Type: 2

Type: U (to select the Unconnected Pins report)

Press: [Alt] O (Ok to print the report)

If nothing is listed in the Unconnected Pins report, all the component pins are connected. If your report shows that some pins are unconnected, connect them now. Return to the Designer Display, delete lines that are not connected properly, and add new lines. Be sure that a small square appears where two lines are connected; otherwise MICRO-LOGIC II may not recognize that they are connected.

When the new drawing is complete, save it and then return to the Unconnected Pins report. Repeat the procedure until all unconnected pins are removed.

Like the Netlist report, you exit the Unconnected Pins report by pressing any key.

Capacity Report

The **Capacity report** lists the usage of each type of component. To generate this report,

Press: [Alt] P

Type: 2

Type: C (to select the Capacity report)

Press: [Alt] O (Ok to print the report)

This information is also provided in the Netlist report. In addition, you can determine the capacity used from the Designer Display screen by pulling down the View menu and selecting option **1:Capacity used**.

To exit the Capacity report, press any key.

Problems

1. Figure 3.11 shows a Netlist report for an unknown circuit. Draw the circuit and identify any potential wiring errors.

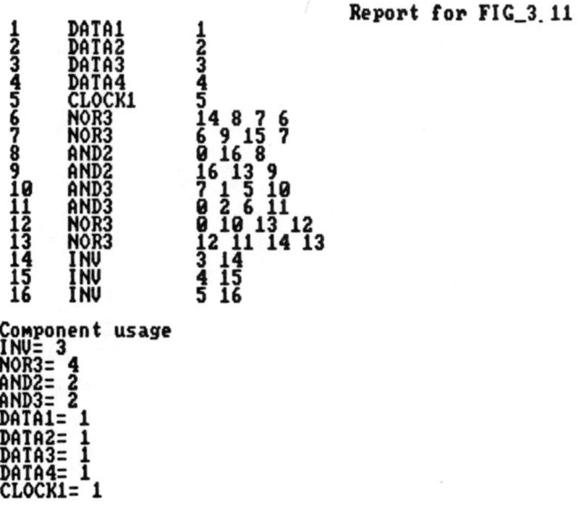

```
                                Report for FIG_3.11
1    DATA1     1
2    DATA2     2
3    DATA3     3
4    DATA4     4
5    CLOCK1    5
6    NOR3      14 8 7 6
7    NOR3      6 9 15 7
8    AND2      0 16 8
9    AND2      16 13 9
10   AND3      7 1 5 10
11   AND3      0 2 6 11
12   NOR3      0 10 13 12
13   NOR3      12 11 14 13
14   INV       3 14
15   INV       4 15
16   INV       5 16

Component usage
INV= 3
NOR3= 4
AND2= 2
AND3= 2
DATA1= 1
DATA2= 1
DATA3= 1
DATA4= 1
CLOCK1= 1
```

Figure 3.11

2. Generate the following input patterns using the Pattern editor. Create the patterns using the fewest possible steps.

Pattern1: 1111111111111111 1111111111111111 0000000000000000 0000000000000000

Pattern2: 1111000011110000 1111000011110000 1111000011110000 1111000011110000

Pattern3: 1111111111111111 0000000000000000 1111111111111111 0000000000000000

Pattern4: 0000111100001111 0000000000000000 0000111100001111 0101010101010101

4 Basic Combinational Circuit Simulation

4 Basic Combinational Circuit Simulation

Tutorial 4 will teach you how to:

- Perform basic circuit simulation of combinational circuits

The Simulator

The MICRO-LOGIC II simulator can simulate the operation of the currently displayed circuit. Let's simulate our FULL_ADD circuit. Load this drawing by pressing [Alt] **F** and typing **2**. In the File dialog box highlight the file FULL_ADD and press [↵].

To simulate the functioning of a circuit, select option **1:Simulate** in the **Simulate** menu.

Press: [Alt] `S`

Type: `1`

Analysis Limits Screen

Your screen should look like Figure 4.1, which shows the **Analysis Limits screen**. Important parameters that regulate the simulation are defined on this screen. Entries on the screen can be edited using the four cursor keys, as well as [←], [Del], [Ins], and [↵].

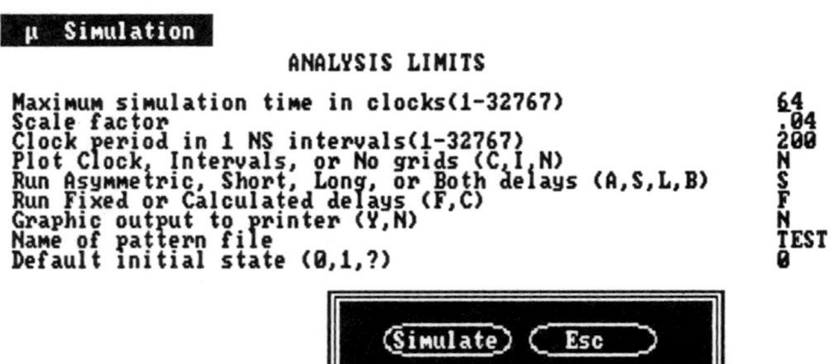

Figure 4.1

Maximum simulation time in clocks (1-32767) specifies how long, in clock cycles, the simulation will last. The default value is 64. Since a full adder can generate only eight different output combinations, let's set the simulation for eight clock cycles. One way to do this is to move the cursor under the 6 in the value 64 just to the right of the entry **Maximum simulation time in clocks**. Then type **8** and press Del.

Scale factor determines the size of the generated output waveforms. Choosing an appropriate scale value for a particular simulation may take some experimentation. Simply pick a value and run the simulation. If the display is not what you expected, change the scale factor and rerun the simulation. For this simulation a scale factor of .33 will make a nice display, so set this value to **.33**.

Name of pattern file defines which Pattern file will be used in the simulation to define the data channels. Since we want to use the BINARY pattern file, edit this entry to **BINARY**. When you're finished, your screen should look like Figure 4.2.

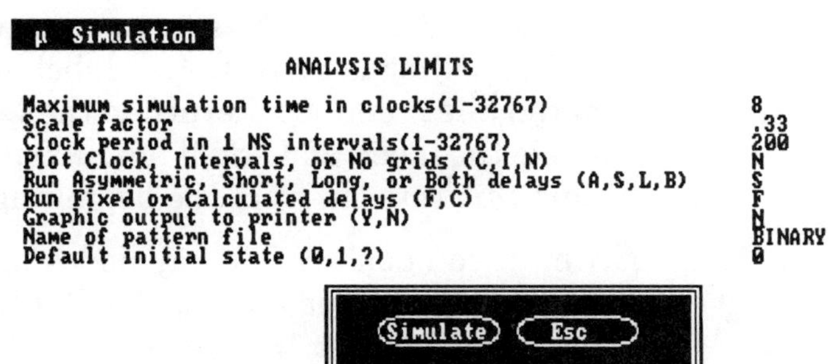

Figure 4.2

Nodes Monitored Screen

Now let's specify the nodes that we want to monitor during the simulation. With the Analysis Limits screen still displayed, select option **E:Edit nodes monitored** in the **Simulation** menu.

Press: [Alt] S

Type: E

The **Nodes Monitored screen** is now displayed, and we can type in the nodes that we want to monitor. We'll monitor all the nodes that were labeled in the drawing, so list all the nodes as shown in Figure 4.3. Use the four cursor keys, as well as [←], [Del], [Ins], and [↵], to edit your entries. Type each entry on a separate line.

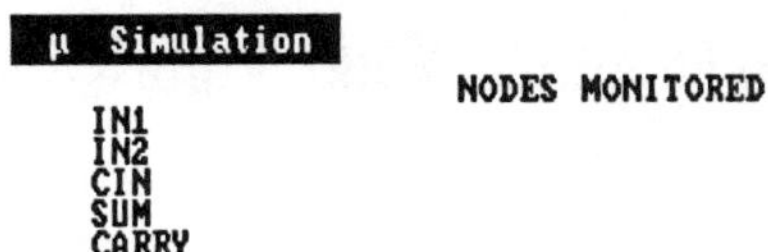

Figure 4.3

Note: After you type a label, such as XX, and press [↵], a dialog box may appear that states, **Can't find XX in label list**. If this happens, type **O** (for Ok) to close the dialog box. Then do the following:

- Check the spelling of your label. If it's incorrect, change it.

The labels entered in the Nodes Monitored screen must exactly match the labels in your drawing.

- If the spelling is correct, your label isn't correctly positioned on the line it defines. Return to the Designer Display screen, delete the label, and add it again. Save the new drawing and start this tutorial from the beginning.

Note: If you don't define any nodes to be monitored by using the Nodes Monitored option, MICRO-LOGIC II will automatically select the first 19 labeled nodes from the drawing.

To exit the Nodes Monitored screen, pull down the **Simulation** menu and choose another option such as **A:Edit analysis values**, which will return you to the Analysis Limits screen. Or you can press [Esc] to return to the screen from which you entered the Nodes Monitored screen—in this case the Analysis Limits screen again. To return to the Designer Display screen, pull down the μ menu and press [Esc].

Simulation

Whether you are still in the Nodes Monitored screen or have returned to the Analysis Limits screen, you can now begin a simulation. Either press [Home], or select option **Home:Simulate** on the Simulation menu.

Mouse A simulation can be started either by clicking option **Home:Simulate** on the **Simulate** menu, or by clicking the Simulate bubble on the Analysis Limits screen.

Your screen should look like Figure 4.4.

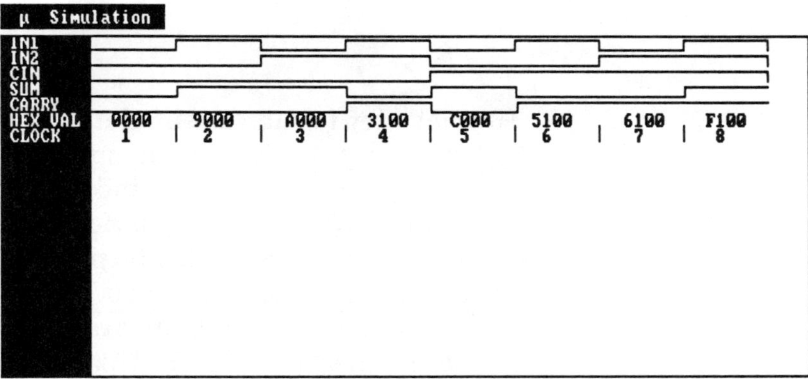

Figure 4.4

This screen displays the results of the simulation run. Down the left side are the labels specified in the Nodes Monitored screen. Note that Clock Cycles 1 through 8 are labeled at the bottom of the display.

The **hex value** labeled just above the clock cycle is the hexadecimal (base 16) value of the displayed waveforms. The top four waveforms make up the most significant digit, with the top waveform being the least significant bit of the most significant digit. The second four waveforms are the second most significant digit, and so on. (Nonexistent waveforms are assumed to have a value of 0.)

For instance, in Figure 4.4, the hex value of the waveforms in the fourth clock cycle is 3100 or, in binary, 0011 0001 0000 0000. Hence the value in Clock Cycle 4 of the first four waveforms displayed from top to bottom are 1100 respectively. This translates to IN1 = 1, IN2 = 1, CIN = 0, and SUM = 0.

Notice that the displayed waveforms correspond to the truth table for the full adder circuit. To verify this, compare the waveforms on the screen with Table 1.1.

The simulation can be run again from this screen by pressing [Home] or by using the Simulate menu.

Now let's experiment with the simulator by trying various options in order to demonstrate some of the simulation features. To do this, we must return to the Analysis Limits screen and redefine some of the parameters.

Press: [Alt] S

Type: A

First let's try to run the simulation for a longer period of time by increasing the maximum simulation time. Using the four cursor keys and [←], [Del], [Ins], and [↵], set the maximum simulation time to 1024. The simulation will now run for 1024 clock cycles. Start the simulation by pressing [Home]. Since the simulation run is quite long, not all of it will fit on the screen at once; hence the simulation results are scrolled across the screen as they are generated. You can pause the scrolling in order to examine some particular point in the simulation by typing **P**. The simulation can be continued by pressing [Esc].

As the simulation continues, notice that a series of commands appears across the top of the screen. We've already used the **P:Pause** command. Try the **E:Expand scale** and **S:Shrink scale** commands by typing **E** and **S**, respectively.

If you didn't get a chance to perform the preceding commands before the simulation ended, start the simulation again by pressing [Home]. If you completed the preceding steps and the simulation is still running, you can end the simulation now by pressing [Esc].

When the simulation has ended (either on its own or after you pressed [Esc]), return to the Analysis Limits screen by pulling down the Simulate menu and selecting the **A:Edit analysis values** option. Reset **maximum simulation time** to **8**. Then set the **scale factor** to **.04** and begin a new simulation by pressing [Home].

Notice that, because the scale factor was reduced, the simulation waveforms are the same but much smaller. The clock cycles and hex values are no longer listed below the waveforms because there is no room for them. Instead, the clock cycles are marked by small vertical lines below the waveforms, and only the last clock cycle displayed on the screen (in this case **C= 8**) is indicated—on the left of the display. The hex value of the last displayed clock cycle (in this case **H=F100**) is also listed on the left.

Simulation Hardcopies

Return to the Analysis Limits screen and reset the **scale factor** to **.33**. Then move the cursor to the entry next to **Graphic output to printer (Y,N)**. The current entry is **N**, meaning that no printer output is generated. If **Y** is entered, then the waveforms will be printed as hardcopy as well as displayed on the screen.

Edit the entry so a printer output will be produced and begin a simulation. A Printer Options dialog box appears just as when a drawing file was printed. Select the option you want by typing **1**, **2**, or **3**. After you make your selection, another dialog box appears as shown in Figure 4.5. Simply press [Alt] **O** (for Ok) to proceed with the simulation. For long simulations, a portion of the waveform can be printed by defining an appropriate **start clock** and **end clock**.

```
µ Simulation
                              ANALYSIS LIMITS
Maximum simulation time in clocks(1-32767)                   8
Scale factor                                                 .33
Clock period in 1 NS intervals(1-32767)                      200
Plot Clock, Intervals, or No grids (C,I,N)                   N
Run Asymmetric, Short, L                                     S
Run Fixed or Calculated                                      F
Graphic output to printe                                     Y
Name of pattern file        Start clock   1                  BINARY
Default initial state (0    End clock     8                  0

                                  Ok
```

Figure 4.5

Note: The Start and End Clock dialog box can produce unexpected results.

The simulation and printing can be ended prematurely by pressing [Esc].

The μ menu functions as it did earlier. Using this menu, or simply pressing Esc, will return you to the screen that activated the current screen.

The **Simulate** menu lists a number of options. We've already examined several, including **Home:Simulate, A:Edit analysis values**, and **E:Edit nodes monitored**. By selecting option **P:Patterns**, you can enter the Pattern editor—a handy feature if during a simulation you want to change some of the input values.

Option **C:Clocks** is discussed in Tutorial 5 in the context of sequential circuit simulation. Options **S:Save run** and **R:Retrieve run** are not available on the student version of MICRO-LOGIC II.

The final option, **T:Text file**, is a toggle option—that is, it can be turned on or off. It is initially toggled off and thus inactive. If it is toggled on, an ASCII text file is created on your data disk when a simulation is performed. The text file, which is created during the simulation run, shows the state of each monitored node at the end of each clock cycle. The filename is FILENAME.VEC, where FILENAME is the name of the drawing being simulated and VEC is an extension automatically added to the filename. This file can be viewed and edited by most word processors. The **T** option will produce a disk file only when a simulation is run.

Note: Simulation values set in the Analysis Limits screen are linked to the drawing file of the circuit being simulated. This fact should be kept in mind for several reasons:

- If you change the analysis limits and then, having returned to the Designer Display, save the logic drawing to disk, the updated analysis limits will also be stored. This means that when a new simulation is begun, the edited version of the Analysis Limits screen will appear.
- If the analysis limits are changed but you don't save the drawing to disk, when you begin a new simulation the old Analysis Limits screen will be produced.

- The File Warning dialog box of Figure 1.9 can appear because either the drawing has changed and has not been saved, and/or the analysis limits have changed and the drawing has not been saved.

Problems

1. Explain the function of the Expand and Shrink commands during a simulation.
2. Assume that after a simulation has been run that the hex value of Clock Cycle 4 is **ab13**. Draw the waveforms for Clock Cycle 4 as they would appear on the screen.
3. Draw a combinational circuit that will implement the following truth table. Then simulate the circuit and verify its correct operation.

A B C D	OUT
0 0 0 0	0
0 0 0 1	0
0 0 1 0	1
0 0 1 1	1
0 1 0 0	1
0 1 0 1	0
0 1 1 0	1
0 1 1 1	0
1 0 0 0	1
1 0 0 1	1
1 0 1 0	1
1 0 1 1	0
1 1 0 0	0

A B C D	OUT
1 1 0 1	0
1 1 1 0	0
1 1 1 1	1

5 Flip-Flops and Sequential Circuits

5 Flip-Flops and Sequential Circuits

In Tutorial 5 you'll learn how to:

- Use flip-flops to make sequential circuits
- Define and use clock signals

Sequential Devices and Circuits

The output of a sequential device or circuit depends not only on the present input but also on some set of past inputs. This set of past inputs is referred to as the present state of the device.

The flip-flop is the basic sequential device used in designing sequential circuits. There are four commonly defined flip-flops: J-K, R-S, D, and T. MICRO-LOGIC II has a number of flip-flops available for use in circuits. You may wish to review the available selection by scanning the Component Selector.

Sequential circuits are divided into two subclasses of circuits: synchronous and asynchronous. Synchronous sequential circuits, used more often, are characterized by a clock signal that regulates the circuit's activities (typically through the use of clocked flip-flops). Hence we must learn how to include clock signals in our circuit drawings and how to use clocked flip-flops.

All of the flip-flops included in MICRO-LOGIC II are clocked flip-flops.

Clock Signals

Clock signals are represented in a logic diagram by a component called a **clock**. MICRO-LOGIC II includes five clock components, labeled CLOCK1, CLOCK2, ..., CLOCK5. Five more clocks can be defined by using the Component editor (see Tutorial 7). In a drawing, clocks are drawn and manipulated just like any other component. A typical clock waveform is shown in Figure 5.1. Notice that only a single clock pulse occurs during a **clock cycle**.

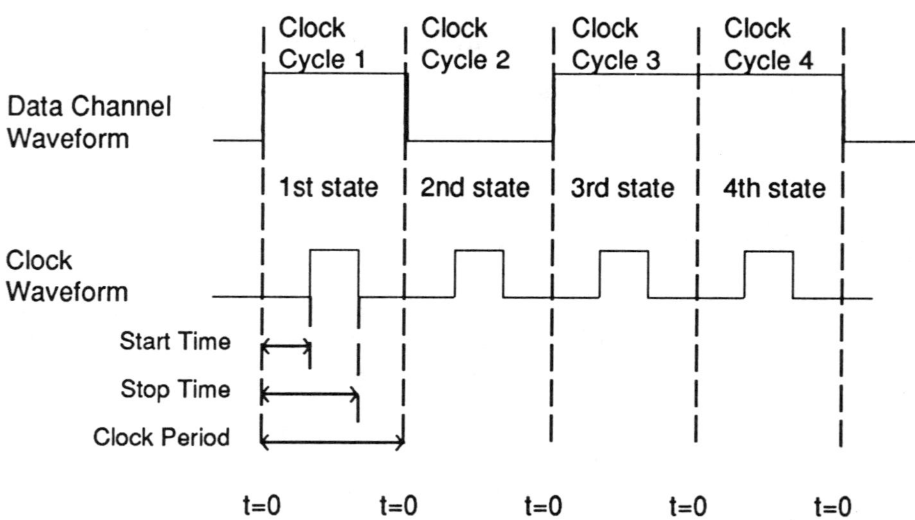

Figure 5.1

Clocks are defined using the Clock editor and the Analysis Limits screen. The length of a clock cycle is defined on the Analysis Limits screen. The Clock editor is used to define the timing of the rising and falling edges of the clock pulse with respect to the start of the clock cycle. To illustrate the function and use of a clock signal, let's use all five clocks and examine their waveforms.

Start with a clean sheet of paper and add the five clocks anywhere in the Display area so long as they don't overlap (see, for example, Figure 5.2). To add the first clock,

Type: `C`

Type: `A`

Type: `CLOCK1`

Press: [↵]

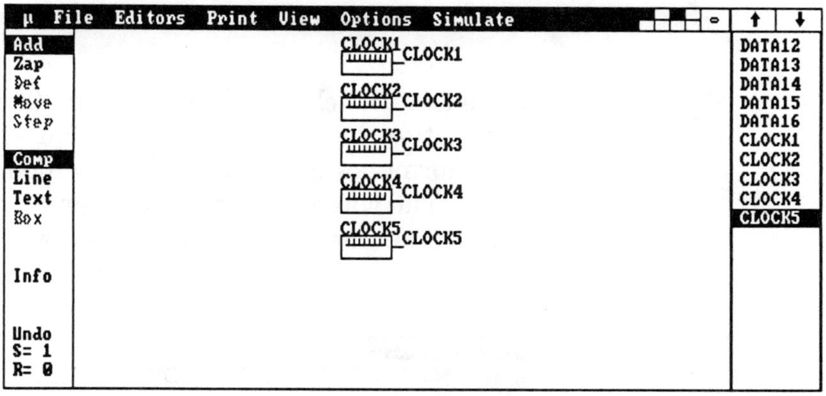

Figure 5.2

Add the other clocks in a similar fashion. Then label the output of each clock so you can identify it during simulation. Label the output of CLOCK1 as **CLOCK1**, the output of CLOCK2 as **CLOCK2**, and so on. Remember that to correctly identify the label with the line, you must position the cursor on the line when the label is added.

Now try to run the simulator:

Press: [Alt] `S`

Type: `1`

Press: [Home]

A dialog box appears indicating that a simulation cannot be performed unless there is at least one data channel in the drawing. Respond to the dialog box by typing **O** for Ok. If the dialog box is still on the screen, return to the Designer Display by pressing [Esc].

Somewhere in the Display area, add a data channel. Return to the Analysis Limits screen by pressing [Alt] **S** and typing **1**. Set up the simulation parameters as shown in Table 5.1.

Maximum simulation time in clocks(1-32767)	2
Scale factor	1

Table 5.1

Note that the clock period is set to 200 nanoseconds. Start the simulation by pressing [Home]. You should see a display of waveforms as in Figure 5.3, in which all five clocks are the same. The waveforms are the same because MICRO-LOGIC II initially defines all the clocks the same.

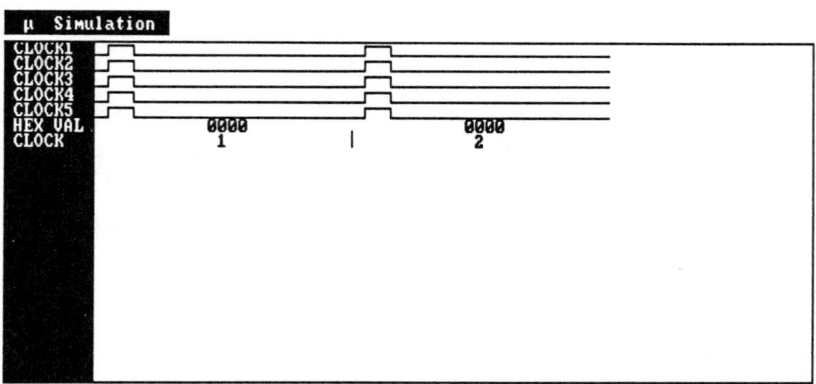

Figure 5.3

Now let's redefine the clocks by using the **Clock editor**. Enter the Clock editor by pressing [Alt] **S** and typing **C**. Your screen should look like Figure 5.4. Remember that although 10 clocks are listed on the screen, only the first five exist as predefined components in the student version of MICRO-LOGIC II. A start and stop time in nanoseconds is defined for each clock. Notice that all five clocks are set with start=10 and stop=30.

This means that all the clock pulses will begin 10 nanoseconds after the start of each clock cycle and will end 30 nanoseconds after the start of each clock cycle.

```
μ                           Clock waveform editor
Name            Start time    Stop time
Clock1          10            30
Clock2          10            30
Clock3          10            30
Clock4          10            30
Clock5          10            30
Clock6          10            30
Clock7          10            30
Clock8          10            30
Clock9          10            30
Clock10         10            30



Esc:Exit editor Tab:change columns
```

Figure 5.4

Verify that the definitions for the five clocks are consistent with the waveforms in Figure 5.3. Remember that for this simulation a clock cycle was defined to be 200 nanoseconds on the Analysis Limits screen.

To edit a start or a stop entry, use the four cursor keys, as well as the ←, Del, Ins, and numeral keys. To move between the start time and stop time columns use the Tab key. Edit the entries as shown in Table 5.2.

	Start	Stop
Clock 1	10	30
Clock 2	40	120
Clock 3	60	120
Clock 4	100	110
Clock 5	180	190

Table 5.2

Now exit the Clock editor by pressing Esc. When the Analysis Limits screen appears, start the simulation by pressing Home.

The waveforms of Figure 5.5 should have been generated. Compare these waveforms with their definitions in Table 5.2.

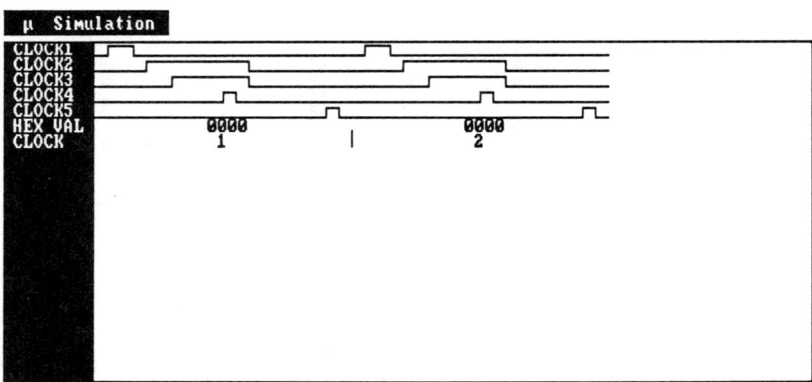

Figure 5.5

Changing the length of the clock cycle is accomplished using the Analysis Limits screen. Return to the Analysis Limits screen by pressing Alt **S** and typing **A**, and redefine the clock period to be 500 nanoseconds. Restart the simulation by pressing Home, and again study the generated waveforms (see Figure 5.6). Compare these waveforms with the set shown in Figure 5.5.

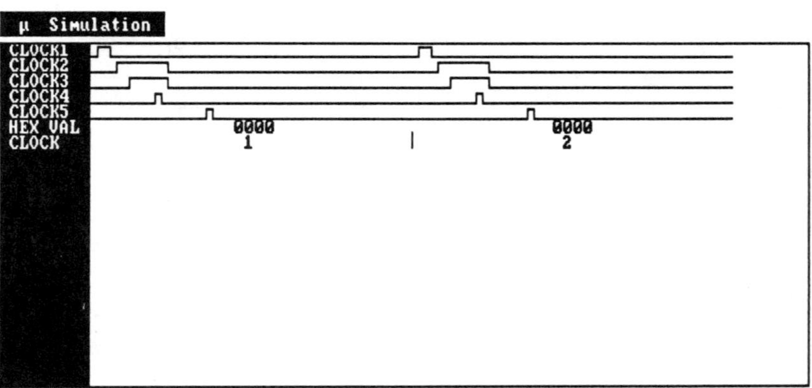

Figure 5.6

Note: Clock definitions (start and stop times) are not stored with a logic drawing like Analysis Limits screen values. Therefore, clock values defined for one drawing will be the same for any other drawing unless they are redefined using the Clock editor.

Although often only one clock signal is needed in a circuit, others are available if needed. In addition, since clock definitions are not stored with a drawing, five different clocks can be defined once in the Clock editor, and then each clock can be used in a different drawing without having to reuse the Clock editor.

A Three-Bit Binary Synchronous Counter

We now have all the tools necessary to draw and simulate a sequential circuit. The circuit we'll use is a **three-bit binary synchronous counter**. We'll use negative edge-triggered J-K flip-flops with clear. The circuit to be simulated is shown in Figure 5.7.

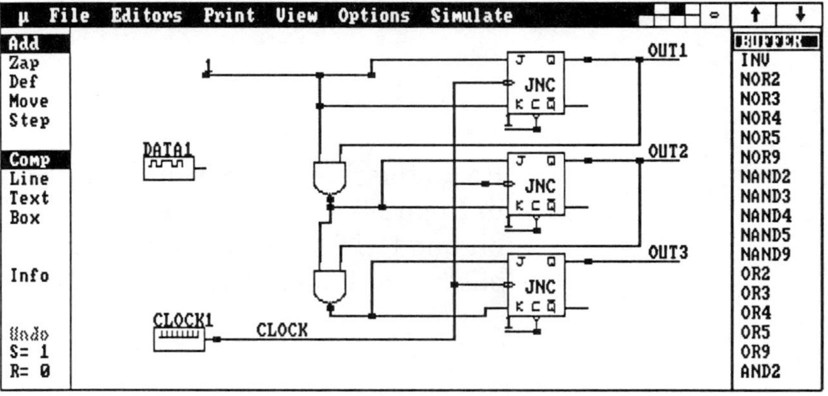

Figure 5.7

First draw the three JNC flip-flops. Then add the two AND2 gates, with the rotation/reflection number set to 1. Then, with the rotation/reflection number set to 0 again, add the CLOCK1 component and the DATA1 component. (Data Channel 1 is not necessary for the circuit to function but is necessary for the simulator to function.)

Next, interconnect the components. CLOCK1 is connected only to the flip-flop clock inputs. The J and K of each flip-flop are connected.

Label the four lines as follows: **OUT1**, **OUT2**, **OUT3**, and **CLOCK**. Finally, label each flip-flop clear input and the unterminated line at the top left of the screen with the label **1**. This label must be positioned at the end of the line.

Note: **1**, **0**, and **?** are also reserved labels when used with unterminated lines (lines with one unconnected end). A line labeled **1** will take on the constant value of 1. A line labeled **0** will take on the constant value of 0. A line labeled **?** will have an indeterminate value. For the labels **1**, **0**, and **?** to function correctly, however, the cursor must be at the very end of the line when the label is added.

Now move to the Analysis Limits screen by pressing [Alt] **S** and typing **1**. Set the analysis limits to the values shown in Table 5.3.

Maximum simulation time in clocks(1-32767)	8
Scale factor	.33

Table 5.3

Begin the simulation by pressing [Home] ; the waveforms shown in Figure 5.8 should appear. With OUT1 the least significant bit and OUT3 the most significant bit, you will notice that the outputs are a binary count from 000 to 111.

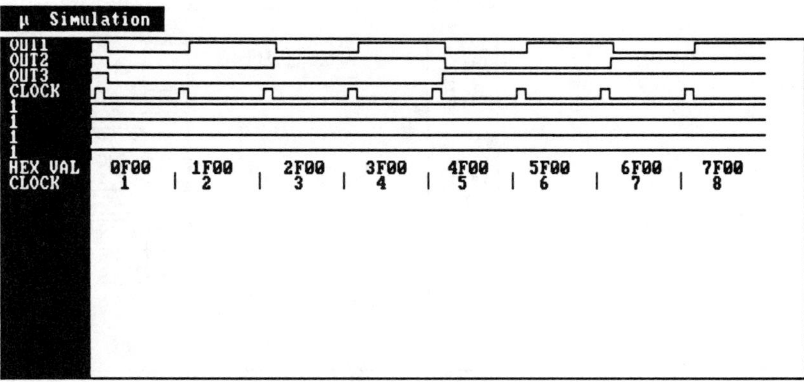

Figure 5.8

Return to the Designer Display screen by pressing Esc and save this drawing by using the File pull-down menu. Save the drawing with the name **BIN_SYNC** (Binary Synchronous Counter).

Problems

1. When a clock is defined, only one start time and one stop time may be defined; hence only one clock pulse will occur during any clock cycle. Is there any way in which two clock pulses can be generated in a clock cycle?
2. Is it possible to define a clock such that a clock pulse extends through more than one clock cycle?
3. Try drawing and simulating a three-bit asynchronous binary counter as shown in Figure 5.9. Does it make any difference whether the input is a clock or a data channel?

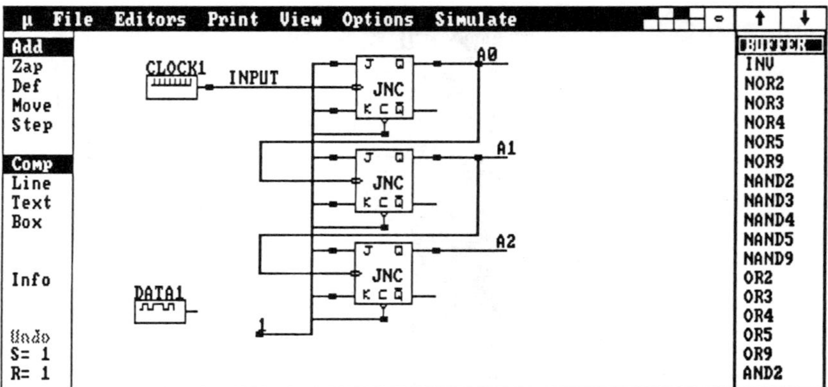

Figure 5.9

4. The circuit of Figure 5.10 detects the binary sequence 101 in the input, X. Whenever this sequence is detected, the output, Z, is set to 1; otherwise the output is 0. For example,

   X = 0 0 1 0 1 1 1 0 1 0 1 1 1 0 1
   Z = _ 0 0 0 1 0 0 0 1 0 1 0 0 0 1
   where _ is an undefined value.

Draw and simulate the circuit to verify its correct operation.

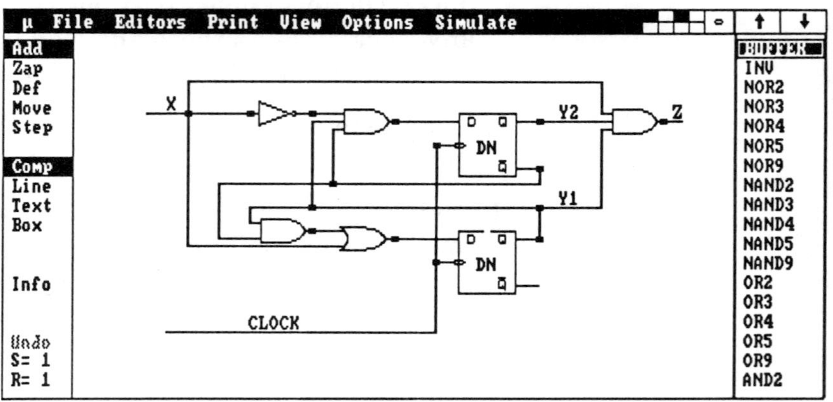

Figure 5.10

6 Additional Simulation Features

6 Additional Simulation Features

Tutorial 6 will teach you how to:

- Enhance the simulation display
- Examine component characteristics
- Interpret the effect of propagation delay

Simulation Waveform Display

Load the **BIN_SYNC** drawing (created in Tutorial 5) by using the File pull-down menu. Enter the simulator by pressing Alt **S** and typing **1**. Edit the analysis limits as shown in Table 6.1.

Maximum simulation time in clocks(1-32767)	4
Scale factor	.6
Clock period in 1 NS intervals(1-32767)	200
Plot Clock, Intervals, or No grids (C,I,N)	N

Table 6.1

Start the simulation by pressing [Home]. Your screen should look like Figure 6.1. Notice that although the beginning of each clock cycle is marked at the bottom of the display with short vertical lines (or grids), no vertical reference grids are drawn in the waveforms themselves. Grids are not marked on the waveforms because you chose the **No grids** option in the analysis limits.

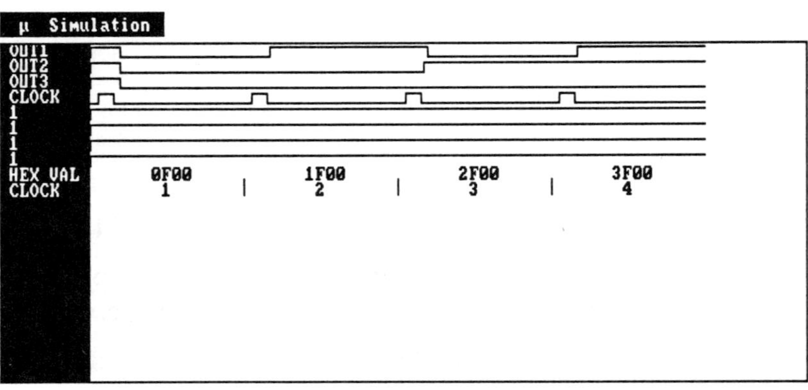

Figure 6.1

Now return to the Analysis Limits screen by pressing [Alt] **S** and typing **A**, and change the **Plot Clock ...** entry to **C** for **Clock grids**. This setting will cause the clock grid line to extend across all the waveforms.

Start the simulation again by pressing [Home] and observe the waveforms. Note that the clock grids now extend across all the waveforms (see Figure 6.2), making the timing of the waveforms easier to identify.

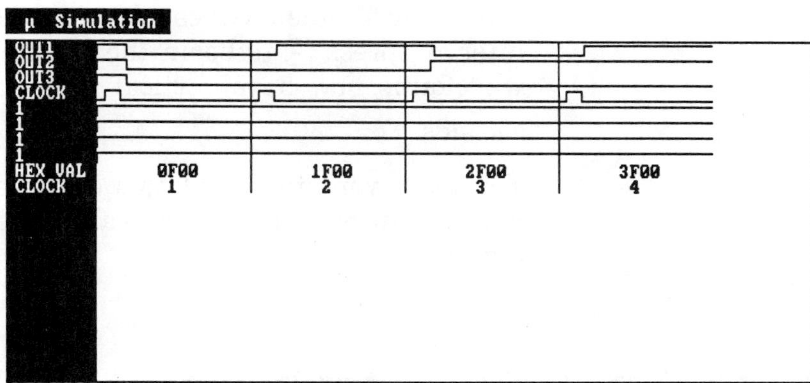

Figure 6.2

Again return to the Analysis Limits screen; change the **Plot Clock ...** entry to **I**, for **Interval grids**. Interval grid lines are similar to clock grid lines but can be defined by the user at any interval, not just on clock cycle boundaries.

Let's set our intervals at 100 nanoseconds (remember that the clock cycle is 200 nanoseconds). First, start the simulation again by pressing [Home]. A dialog box now appears requesting a marker interval value. Although a value is already entered in the dialog box, change this value to 100 and press [↵].

The waveforms in your display should now look like Figure 6.3, in which a vertical grid mark is placed every 100 nanoseconds.

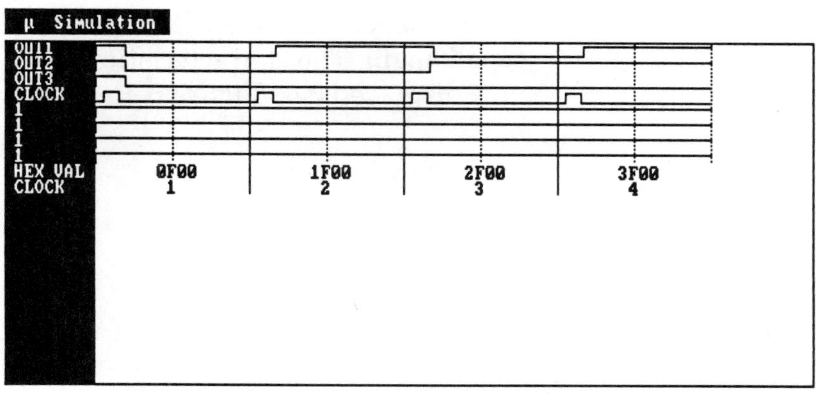

Figure 6.3

Start another simulation by pressing [Home] : set the marker interval to **50** when requested to do so. After setting the marker interval, press [↵] to begin the simulation. Observe that the waveforms now have vertical grid markers every 50 nanoseconds.

You may want to repeat this exercise using some different marker intervals. Some interval values are more helpful than others.

Component Characteristics and Library Information

Return to the Designer Display screen and use the File menu to load the **FULL_ADD** drawing. You may save the BIN_SYNC drawing again if you wish (remember that the analysis limits were changed).

The FULL_ADD drawing contains four different kinds of components: AND2 gates, EXOR gates, data channels, and an OR2 gate. Each component is defined by certain specific characteristics or parameters. These characteristics, stored in a **library**, can be observed by activating the **Info** function on the Function Selector. When the cursor is moved to a component and the **Info** function is selected, typing **I** will display that component's library information on the screen.

Move the cursor to Data Channel 1 and type **I**. Notice that the **Info** function has been selected in the Function Selector. Now type **I** again. The characteristics of the component DATA1 are now displayed (see Figure 6.4).

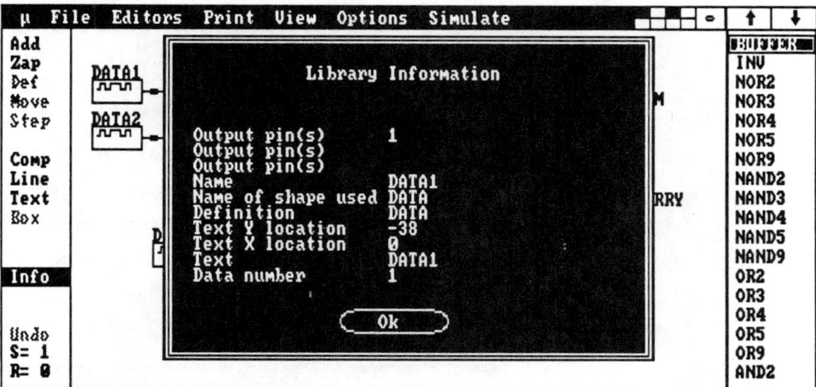

Figure 6.4

The displayed information lists the following characteristics about the component in general (and DATA1 specifically):

- A listing of those pins that are outputs (only pin 1).
- The name of the component as found in the Component Selector (DATA1).
- The name of the shape used in the Designer Display (DATA shape).
- The definition of the function implemented (DATA or data channel).
- A label and its position in the shape (Y = -38, X = 0, text = DATA1).
- The number of the data pattern to be used to define this data channel (pattern 1).

Notice that the label, DATA1, is positioned at location X = 0 and Y = -38. X and Y are Cartesian coordinate values with respect to the origin of the component shape. You may have noticed that whenever a component is added to the Display area, the component is always drawn in the same position with respect to the cursor. The position of the cursor when the component is drawn indicates the origin of the symbol. For instance, when a data channel is drawn, the cursor is always in the upper left corner of the box in the data channel symbol. Hence the text DATA1 is drawn in the same position as the origin on the X axis, but is -38 units above the origin on the Y axis.

Note: The size of these units, the creation of the component shapes, and the placement of the shape origin are defined in the Shape editor. The Shape editor is not available in the student version of MICRO-LOGIC II.

Now return to the Designer Display by typing **O** (for Ok), and display the characteristics of either EXOR gate by moving the cursor over one of the gates and typing **I**. The information displayed should look like Figure 6.5.

Figure 6.5

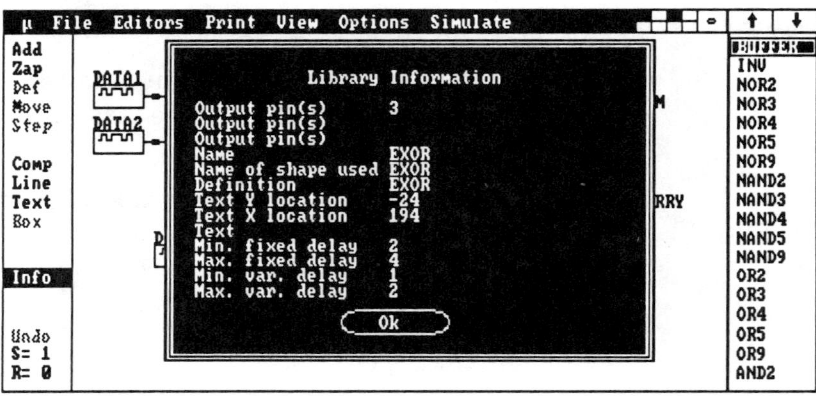

Figure 6.5 shows the same kind of information as Figure 6.4, with two exceptions:

- Since this is not a data channel, no pattern is identified with the component.
- Four propagation delay values are listed (Min. fixed=2, Max. fixed=4, Min. var.=1, Max. var.=2).

Now examine the rest of the displayed information. You may wish to check the library information on the other data channels or other logic gates.

Every digital device has a propagation delay (gate delay). That is, after an input signal is changed, there is a certain delay before that change is reflected on the output (see Figure 6.6). In MICRO-LOGIC II these delays are defined for all Boolean components (AND, NAND, OR, NOR, EXOR, invertor, and buffer). Data channels and clocks do not have propagation delays. Flip-flop delays will be discussed in Tutorial 7 in the context of macros.

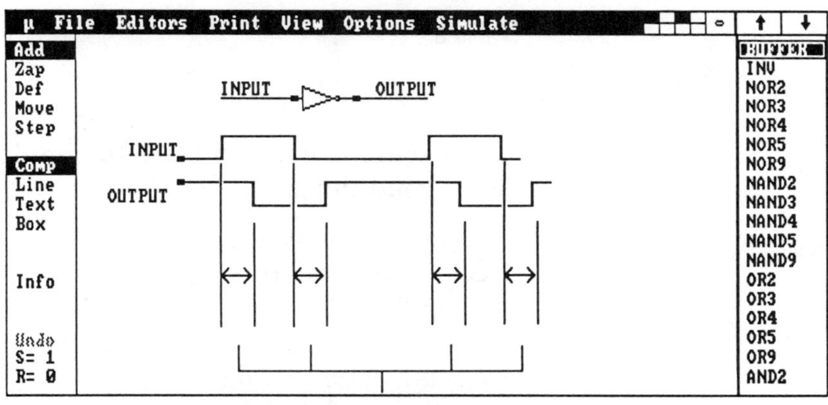

Figure 6.6

Propagation Delay

Four delays are defined for each Boolean element: a minimum and maximum fixed delay, and a minimum and maximum variable delay. These values, or some combination of them, are used during simulation to determine the simulated propagation delay. The particular combination of values to be used is determined by a setting on the Analysis Limits screen.

With the **FULL_ADD** drawing displayed, enter the simulator by pressing [Alt] **S** and typing **1**. Set the analysis limits as shown in Table 6.2.

Maximum simulation time in clocks(1-32767)	2
Scale factor	10
Clock period in 1 NS intervals(1-32767)	200
Plot Clock, Intervals, or No grids (C,I,N)	I
Run Asymmetric, Short, Long, or Both delays (A,S,L,B)	S
Run Fixed or Calculated delays (F,C)	F
Graphic output to printer (Y,N)	N

Table 6.2

Eight options are available to simulate various propagation delay conditions:

- Asymmetric fixed delay
- Short fixed delay
- Long fixed delay
- Both short and long fixed delays
- Asymmetric calculated delay
- Short calculated delay
- Long calculated delay
- Both short and long calculated delays

An option is chosen by entering the appropriate characters on the Analysis Limits screen (A, S, L, B, F, C).

If you choose a **fixed delay**, the minimum and/or maximum fixed delay values are used for each component (see Figure 6.5). The propagation delay of the components will remain the same—regardless of how they are used in the circuit.

In a real circuit, however, propagation delay varies depending on the number of other gates that are driven by the output of that component. The **calculated delay** option simulates this effect. If you choose the calculated delay option, the minimum and maximum values of delay are as follows:

Minimum Delay = ***Min. fixed*** + (***Min. var.*** x ***Fanout***)

Maximum Delay = ***Max. fixed*** + (***Max. var.*** x ***Fanout***)

where ***Min. fixed***, ***Min. var.***, ***Max. fixed***, and ***Max. var***. are the defined values of the component as found in the library, and ***Fanout*** is the number of gate inputs driven by the output of that component.

Therefore a minimum delay and a maximum delay are determined by your choosing either a fixed delay or a calculated delay. Selecting an asymmetric delay, a short delay, a long delay, or both short and long delays will determine whether a minimum delay, a maximum delay, or some combination of both is used.

If you select **short**, only minimum delay values are used. If you select **long**, only maximum delay values are used.

If you select **asymmetric**, the minimum delay value is used when the output changes from 1 to 0; the maximum delay value, when the output changes from 0 to 1.

If you select **both**, both minimum and maximum values are used. This produces an uncertainty interval that begins at the minimum delay and ends at the maximum delay. Uncertainty intervals are denoted by waveform values of both 1 and 0, producing a solid box-like output waveform.

We'll examine only the fixed short, fixed long, and fixed both options in this tutorial. It is suggested that you try all the different options yourself (see Problem 6.2).

Since the **fixed short** delay option is already set, let's look at it first. Start a simulation by pressing [Home]. When asked for a marker interval, press [↵]. (If MICRO-LOGIC II won't accept the current value, change it to 13.3333.) The waveforms should look like Figure 6.7. Observe the **SUM** signal at the point at which it changes state from 0 to 1 (just after the second clock cycle begins). Notice that the transition occurs closer to the clock cycle grid than to the first interval marker grid to its right.

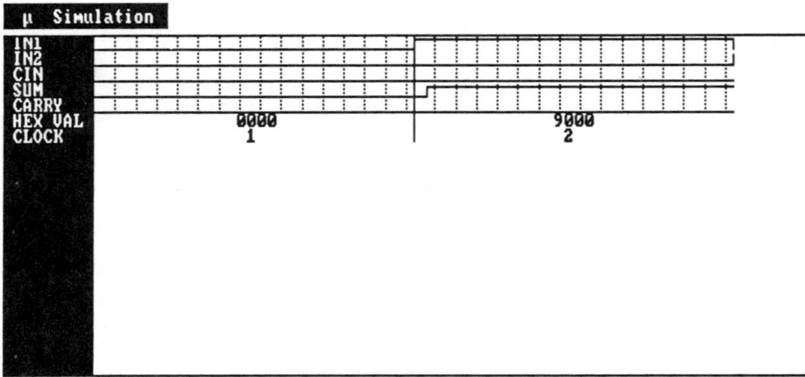

Figure 6.7

Now see what happens if you select the **fixed long** delay option. Return to the Analysis Limits screen. Change the entry at **Run asymmetric ...** to L for **long**. Start the simulation by pressing [Home] and select the marker interval by pressing [↵]. The waveforms should now look like Figure 6.8. The **SUM** signal still changes state from 0 to 1 just after the second clock cycle begins. Notice, however, that the transition now occurs closer to the interval marker grid on its right than to the clock cycle.

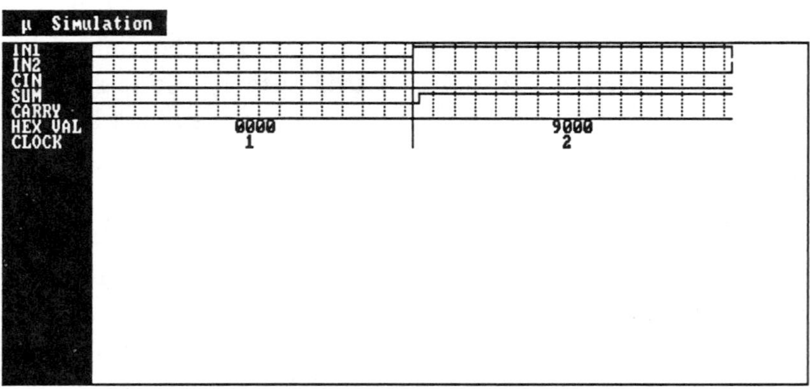

Figure 6.8

Finally, let's see the effect of selecting the **fixed both** delay option. Return to the Analysis Limits screen. Change the entry at **Run asymmetric ...** to **B** (for **both**). Start the simulation by pressing [Home] and again select the marker interval 13.3333 by pressing [↵]. The waveforms now look like Figure 6.9. Notice the box in the **SUM** waveform at the transition from 0 to 1. This represents uncertainty in the transition, the uncertainty being the difference between the minimum and maximum delay.

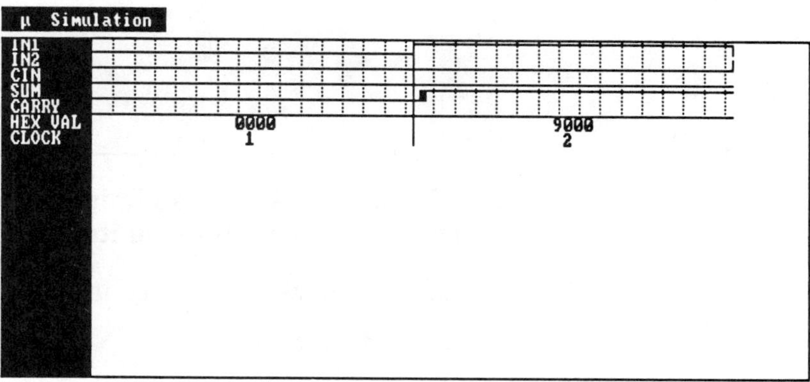

Figure 6.9

Default Initial States

Before a simulation can begin, the simulation software must first initialize the states of all lines in the circuit. Unterminated lines are given a default value, specified by an entry on the Analysis Limits screen. (An unterminated line is a line in which one end of the line is not properly connected to a component or another line, or not properly labeled with **1**, **0**, or **?**.) The default value is either 0, 1, or ? (uncertain). Let's design a small circuit to demonstrate the effect of the default initial state.

Begin with a clean piece of paper. Draw the circuit shown in Figure 6.10. Notice that one input on each of the two gates is left unconnected.

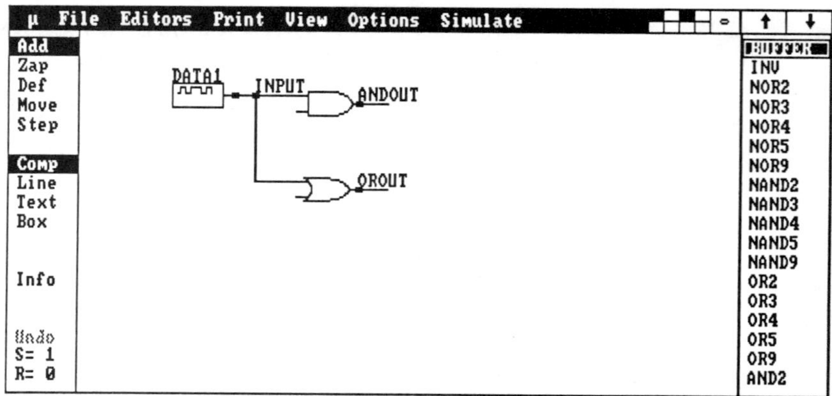

Figure 6.10

Now display the Analysis Limits screen by pressing Alt **S** and typing 1. Set the analysis limits as shown in Table 6.3.

Maximum simulation time in clocks(1-32767)	4
Scale factor	.6
Clock period in 1 NS intervals(1-32767)	200
Plot Clock, Intervals, or No grids (C,I,N)	I
Run Asymmetric, Short, Long, or Both delays (A,S,L,B)	S
Run Fixed or Calculated delays (F,C)	F
Graphic output to printer (Y,N)	N
Name of pattern file	BINARY
Default initial state (0,1,?)	0

Table 6.3

Start a simulation by pressing Home. Since the default initial state is 0, both unconnected lines in the circuit are set to 0. As expected, the output of the AND gate is a constant 0, while the output of the OR gate follows the signal, INPUT.

Now return to the Analysis Limits screen and change the **Default initial state ...** to **1**. Repeat the simulation by pressing [Home] and observe the outputs. Notice that the unconnected lines have now been set to **1** and that, as expected, the output of the AND follows the signal, INPUT, while the output of the OR is a constant 1.

Again return to the Analysis Limits screen and change the **Default initial state ...** to **?**. Repeat the simulation by pressing [Home]. The unconnected lines are now uncertain. Compare the waveforms with those you would expect to see under these conditions.

We won't use this drawing again in the tutorials, but you may want to save it in order to test other analysis values.

Unstable Initial Circuits

When a simulation begins, the simulator initially attempts to evaluate all signal lines so that a stable circuit condition is reached. Occasionally a stable circuit condition cannot be reached, especially in circuits containing flip-flops or other feedback paths. This is usually because of flip-flops in the circuit that are not preset or not reset, or because of similar networks that contain feedback loops without a loop-breaking reset pin.

If a stable circuit condition is not reached, a dialog box appears on the screen indicating that initial states are being evaluated. Press [Esc] to exit the initial state evaluator and begin the simulation even though a stable state has not been reached.

Problems

1. Draw the circuit of Figure 6.11. Simulate the circuit with the following limits set:

Maximum simulation time in clocks(1-32767)	6
Scale factor	.33

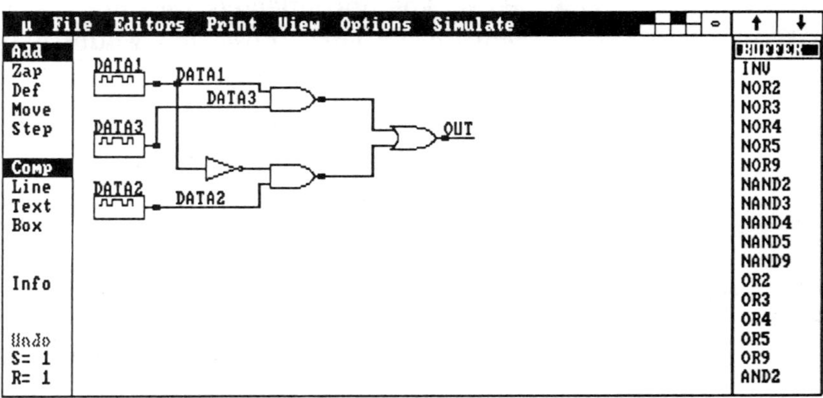

Figure 6.11

Leave all other limits unchanged from their default state. Notice that a small "glitch" appears in the OUT waveform as the fifth clock cycle begins. This glitch indicates the presence of a static hazard in the circuit. The output should remain a constant 1 when the inputs change from DATA 1,2,3 = 1,0,1 to 0,1,1.

a. Verify that this "glitch" should not appear by constructing a truth table for this circuit.

b. Explain why the glitch occurs.

c. Suggest a way to eliminate the glitch.

2. Suggest a method for demonstrating the effect of using a calculated delay as opposed to a fixed delay. Demonstrate your method to verify that it works.

3. Calculate the actual difference between the short fixed delay and the long fixed delay in the SUM of the full adder examined in this tutorial.

7 Macros, New Components, and Tri-State Buses

7 Macros, New Components, and Tri-State Buses

In Tutorial 7 you'll learn how to:

- Create and use macros
- Edit existing component characteristics
- Create new components
- Simulate tri-state gates and buses

Macros

Basic components in MICRO-LOGIC II consist of the Boolean gates AND, NAND, OR, NOR, invertor, buffer, and EXOR, as well as data channels and clocks. More complex components, such as flip-flops, are created by using these basic components to create more complex circuits. These circuits are stored as **macros**.

A macro can be thought of as a program that you can create, save, and then use whenever you wish instead of having to create it again. MICRO-LOGIC II uses macros for complex circuits. Each newly defined macro becomes a component that can be used like any other component.

A macro component does differ from a basic component in some ways. For example, propagation delays are not specifically defined for a macro component. Rather, propagation delay in a macro is determined from the delays in the defining circuit.

Let's examine two existing macros and then define our own new macro. The first macro we'll consider is the negative edge-triggered J-K flip-flop with clear input.

First prepare a clean sheet of drafting paper on the screen:

Press: [Alt] F

Type: 5

Use the cursor keys to move the cursor to the center of the display area. Add the component **JNC** to the paper by selecting the **Comp** and **Add** functions in the Function Selector, and typing **JNC** in the dialog box.

Figure 7.1 shows the JNC component, which should now be in the Display area. This component is a macro (as are all flip-flops). That is, the JNC component is defined by a circuit consisting of a number of simpler components. The circuit that defines the operation of this flip-flop is stored as a macro. When you use the JNC component (the small box in the Display area), you are in effect using another circuit. We can easily examine the circuit that defines this macro component because it is stored as a drawing with the name **JNC**. Now load the drawing labeled **JNC**.

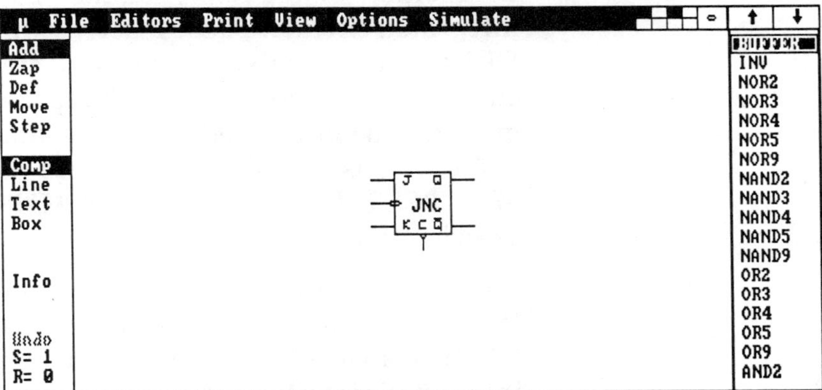

Figure 7.1

Figure 7.2 shows the circuit that defines the macro JNC. Notice that one of the advantages of using a macro is that it reduces drawing clutter. Macros allow us to define and draw complex functions using only a very simple graphics shape. Another advantage of macros is that you can nest them by defining a macro that contains another macro or macros.

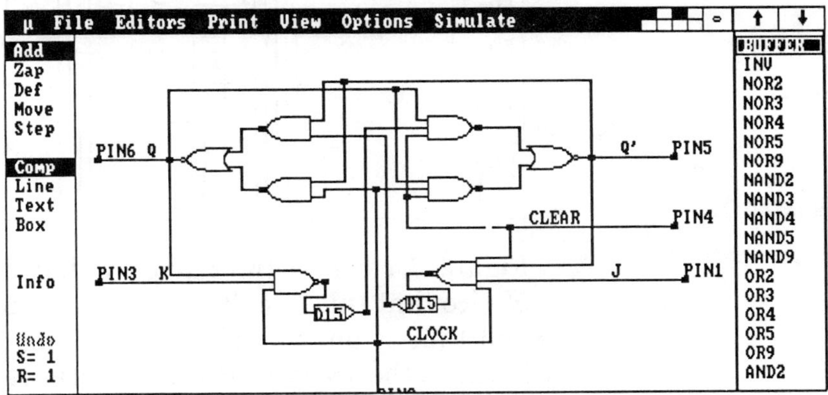

Figure 7.2

Warning: You must be very careful not to change or erase a drawing that defines a predefined macro. Changing or erasing the drawing can cause the defined macro to work improperly or not at all. If you do accidentally change or erase a drawing that defines a macro, it must be reloaded from the original data disk supplied with MICRO-LOGIC II.

In addition to flip-flops, another set of macros consists of integrated circuits. These macros, which are represented by a DIP package symbol, implement standard TTL integrated circuits. The macro labeled 175 represents a 74175 package, a quad D-type integrated circuit. A sample circuit containing a 175 device is included on the data disk in a drawing called **MACTEST**. Load the drawing **MACTEST**. (Do not resave the drawing **JNC**.)

Figure 7.3 shows what you should see in the Display area. In the center of the drawing is a 16-pin integrated circuit package (DIP16) labeled 175. This is a macro representing the 74175 integrated circuit. To examine the circuit that defines the 175 macro, load the drawing labeled **175**.

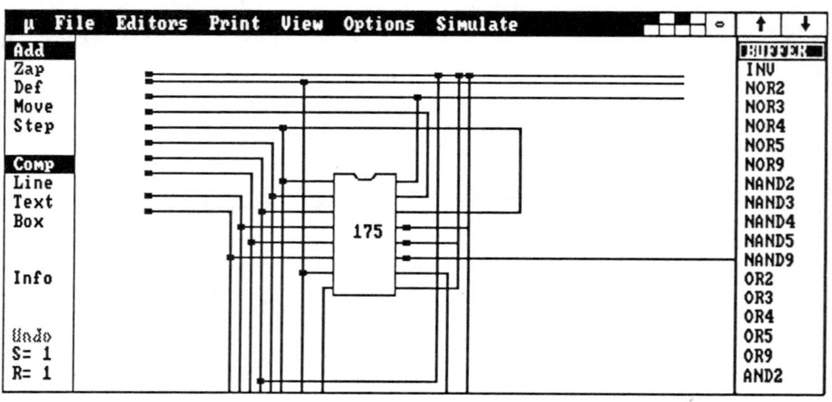

Figure 7.3

Figure 7.4 shows the circuitry that defines the 175 macro. Notice that this circuit also contains macros—DNC macros—and thus is an example of nested macros. Notice the **PINm** labels in the drawing. PINm labels are special labels that make this drawing a macro drawing. They are discussed in the section Creating/Defining Macros later in this tutorial.

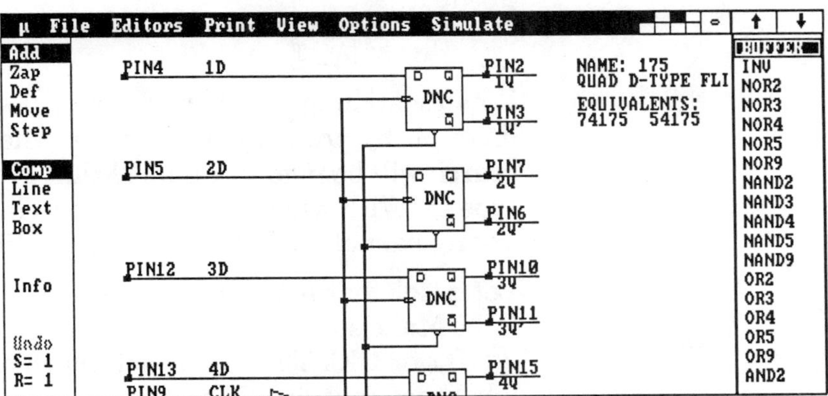

Figure 7.4

In summary, the important difference between a macro component and a basic component lies in the way a macro is defined. Once a macro is defined, however, it is used in a drawing like any other component.

View Search

Before learning how to define a new macro, let's examine a feature of the Designer Display that helps you update large drawings.

You may wish to change a particular element in a drawing. In a large drawing, you must move the paper to locate the element. To move the paper and find the element quickly, use option **7:Search** in the View menu. The Search option allows you to search for either a component or specific piece of text in a drawing.

For example, with the macro drawing **175** still displayed, pull down the View menu by pressing Alt **V**. Then select the Search option by typing **7**. The dialog box that appears allows you to specify a search for either a component or for text. For a component,

Press: Alt G (for Gate)

For text,

Press: [Alt] T

When searching for text you can toggle the **Whole word** option on or off by pressing [Alt] **W**. With whole word on (checked), a search will be made for a label that is exactly like the text you type in. With whole word off, a search will be made for a label or any part of a label that is exactly like the text you type in.

Let 's look for the text **PIN** in the drawing. Press [Alt] **T** and type **PIN**. Make sure that whole word is off (not checked) and press [Alt] **O** (for Ok). The cursor will move to the first instance of a label that contains **PIN**.

Pull down the View menu and select the Search option again. Toggle whole word on (make sure it's checked) and press [Alt] **O** again. This time a dialog box will indicate that a match was not found. No label exists that consists only and exactly of the three characters PIN.

Press [Alt] **O** and again select the Search option of the View menu. Select the Gate option by pressing [Alt] **G**. Now type **INV** in order to search for an invertor. Activate the search by pressing [Alt] **O** or by pressing [↵]. The cursor moves to the first invertor in the drawing.

Try searching for other text and components in the drawing. After you finish, create a clean sheet of paper. (Don't resave the 175 drawing.)

Creating/Defining Macros

You can create your own macros. For example, you can expand the number of available integrated circuit packages. Let's do this now by defining a 7482 package (a two-bit binary full adder). First, however, let's review the difference between a regular drawing and a drawing that defines a macro.

A macro drawing is a drawing that includes text labels that define the connections with the pins of the shape it is represented by. The text labels PIN1, PIN2, PIN3...PIN40—referred to as **PINm** text labels—are reserved for this purpose. A drawing that includes PINm labels is stored as a macro by MICROLOGIC II.

Defining a 7482

We could design a two-bit full adder by using two of our FULLADDER circuits, but instead let's use a slightly faster circuit, shown in Figure 7.5. Figure 7.6 shows the pinout for the 7482 integrated circuit package.

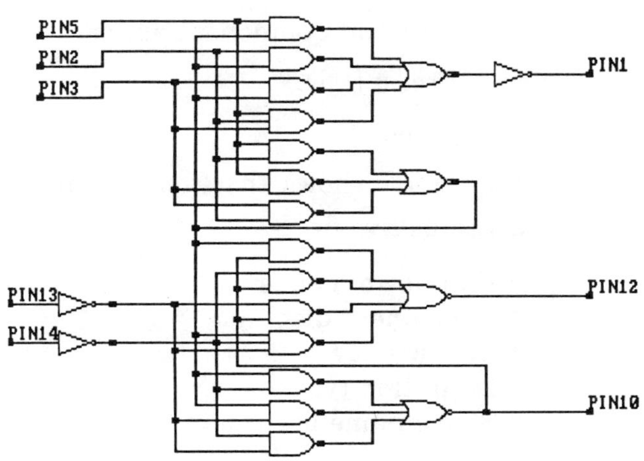

Figure 7.5

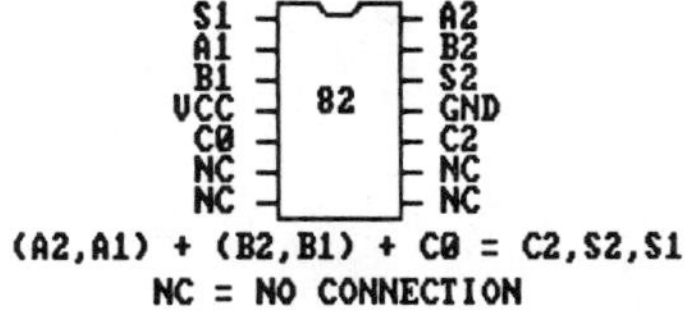

Figure 7.6

Draw the circuit as shown in Figure 7.5. Note that the entire drawing cannot fit on the screen at once. Carefully label the inputs and outputs as shown. The definition of the inputs and outputs can be determined by noting the pin number of the signal on Figure 7.5 and comparing it with the function of the pin as described on Figure 7.6. After your drawing is completed, save it under filename **82**, for a 7482 two-bit full adder.

In order to use your newly defined macro, first you must create a new component.

Component Creation Using the Component Editor

The **Component editor** is used either to change the definition of existing components or to create new components (or macros). Use care in changing existing components, since changes affect not only the performance of that component but also the performance of any macro that uses that component. In general, you need not change an existing component.

To access the Component editor,

Press: [Alt] E

Type: `1`

Figure 7.7 shows the Component editor screen. The screen contains five fields:

- Pull-down menus
- Component display
- Edit window
- Definition types
- Shape name list

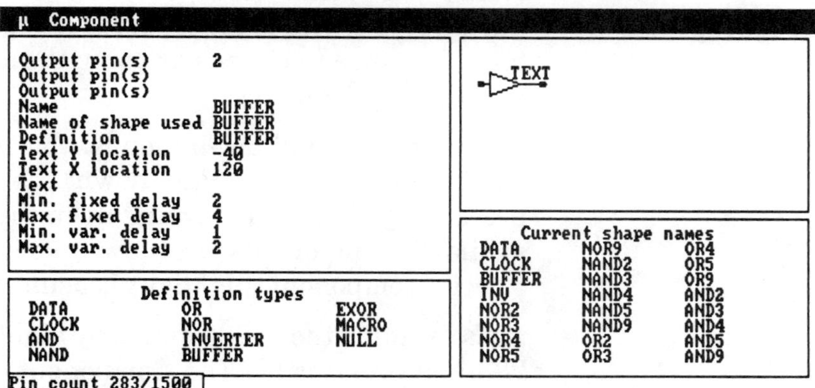

Figure 7.7

The **Definition types** field, in the lower left corner of the screen, lists the five possible definitions for a component: data, clock, Boolean, macro, and null. Each component must be defined as one of these five types.

A data component is a data channel, a clock component is a clock, and a macro is a macro. A Boolean component is one of the Boolean functions: AND, NAND, OR, NOR, invertor, buffer, or EXOR. A null component is a component with no defined inputs or outputs, and thus not yet defined for use. It cannot be used in a drawing and exists merely to allow for the addition of new components by being converted into another type of component.

The **Current shape names** field, in the lower right corner of the screen, lists those shapes that can be used as the graphics symbol to represent the component. There is also a second screen of shape names, which can be accessed by pressing ⊟. Examine both screens to familiarize yourself with the available shape names. The meanings of the shape names are self-explanatory.

The **Component display** field, in the upper right corner of the screen, displays the shape selected for the component and the location of the text field. Inputs are marked with a square and outputs with a circle.

The **Edit window**, in the upper left corner of the screen, allows you to make or to change entries that define a component.

Notice that this window contains the same information displayed by the **Info** function on the Designer Display (see Figure 6.4). You can edit entries by using the four cursor keys, as well as ←, Del, Ins, and ↵.

There are two pull-down menus: the standard μ menu and a **Component menu**. Pull down the Component menu by pressing Alt **C** and then typing **S** (for the **Select** option). All the available components are now displayed on the screen. To select a component, highlight it using the cursor keys.

Let's examine the 74283 macro by highlighting the component **283** and pressing ↵. The Component editor will now display the characteristics for the 283 macro.

Pull down the Component menu again and type **S**. If you move the cursor all the way to the right margin, you'll note that a blank entry is highlighted. This is a null entry (or component), and you can use it to define new components, or macros.

You can access a second page of null entries by pressing Alt **N** or by pressing PgUp. Examine the next page and then return to the first page of components by pressing Alt **N** or PgUp. Highlight **DATA7** by using the cursor keys and then press ↵. The definition of DATA7 is displayed.

The Component menu includes three other functions: **Find**, **Move**, and **Hardcopy**. Find is not available in the student version. The Move function will move a component both in the Component Selector and in the Component editor list, a desirable feature if you use some components more frequently than others. The Hardcopy function displays the complete list of components and their characteristics on the screen or prints them on a printer.

Now let's complete the definition of our macro. Pull down the Component menu by pressing Alt **C** and activate the Select function by typing **S**. Move the cursor to the first null entry at the top of the last column, which is empty, on the first page of the Select Component display and press ↵. A null component definition is now displayed. Edit the entries in the Edit window as shown in Table 7.1 using the four cursor keys, as well as ←,

Del, Ins, and ↵. When using the Edit window, define the **Name**, **Name of shape used**, and **Definition** before defining the other entries.

Output pin(s)	1,10,12
Output pin(s)	
Output pin(s)	
Name	82
Name of shape used	DIP14
Definition	MACRO
Text Y location	68
Text X location	118
Text	82

Table 7.1

After completing the definitions, pull down the Component menu and activate Select. You'll notice that the new component 82 is now listed in the display and is available for use. When you exit the Component editor, this new component definition is saved on disk and appears in the Component Selector along with the other components. Exit Select by pressing Esc.

Before leaving the Component editor, position your new component in numerical order with the rest of the integrated circuit components. Pull down the Component menu and select the **Move** option by typing **M**. Use the cursor keys to select the component to be moved, in this case the new component, 82, and press ↵. Then use the cursor keys to select the component that 82 will be placed after—in this case 74—and press ↵. You can check that the move was made properly by using the Component menu and activating the Select command.

In addition to creating new macro components, additional clocks and data channels can be defined: **CLOCK6, CLOCK7,**

..., **CLOCK10** and **DATA17, DATA18, ..., DATA32**. If you wish to define additional components, the best way to do it is by copying the characteristics of an existing clock or data channel, changing only the name and label of the component.

After you finish, exit the Component editor by pressing [Esc].

Nested Macros

Nested macros can be created to any depth desired as long as the total system capacity for components, lines, nodes, and so on are not exceeded (see the Reference Section). Macros should be drawn in order from the lowest-level macro to the highest-level macro. If any macros are drawn out of order or if a lower-level macro is later changed, the **Update Macros** command must be executed in order for the higher-level macros to use the lower-level macros. The Update Macros command is found in the File menu on the Designer Display screen.

Wired-OR Connections

Although a single output can be connected to several different inputs, generally two different outputs cannot be connected. One exception to this rule is when open collector gates are used. Open collector gates can have their outputs connected, creating a **wired logic** connection. With MICRO-LOGIC II **wired-OR** and **wired-NOR** connections can be made.

Creating a Wired-OR Connection

Pull down the Options menu and make sure that option **2: Wired-Or** is selected. Then, starting with a clean piece of paper, draw the three-buffer circuit shown in Figure 7.8. The connection of the output of the two left-hand buffers creates a wired-OR connection in MICRO-LOGIC II.

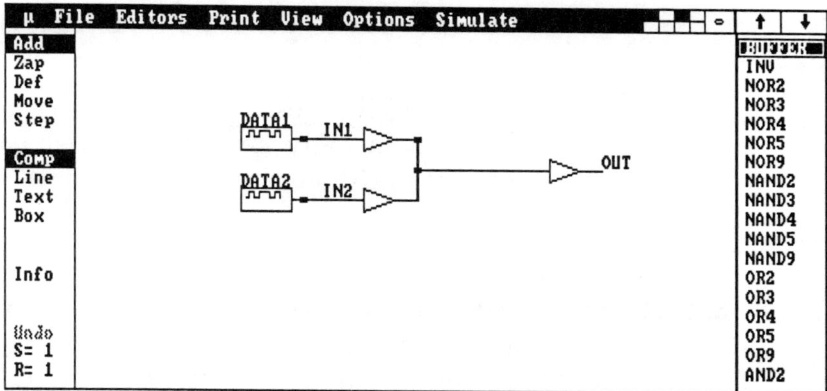

Figure 7.8

Now let's simulate operation of the circuit in order to examine the effect of the wired-OR configuration. Enter the Simulator, and on the Analysis Limits screen set the limits as shown in Table 7.2.

Maximum simulation time in clocks(1-32767)	8
Scale factor	.33
Name of pattern file	BINARY

Table 7.2

Leaving all other limits at their preset values, start the simulation by pressing [Home]. Observe the waveforms and note that **OUT** is 1 if either **IN1** or **IN2**, or both, are 1. **OUT** is 0 only when both **IN1** and **IN2** are 0. This is the **OR** function.

Exit the simulation by pressing [Esc]. Generate a Netlist report by opening the Print menu and selecting **2:Print Reports**. Make sure the options are correct, and then type **O**. Notice that an additional component has been added to the circuit—a **wired-OR** gate. Hence as far as the simulation is concerned, connecting the outputs of gates is the same as inputting the gate outputs into an OR gate.

If desired, a wired-NOR connection can be made instead of a wired-OR. A wired-NOR connection will be made if option

1:Wired-Nor on the **Options** menu is activated. Only one of the two options, wired-OR and wired-NOR, can be active at any one time.

Return to the Designer Display screen from the Netlist report by pressing any key. If you wish you may first save this drawing, although we won't use it again.

Tri-State Buses/Devices

Although wired-OR connections aren't used very often, they have a special use in MICRO-LOGIC II. No tri-state components are available in MICRO-LOGIC II, but these components can be simulated with a wired logic connection and some additional logic.

We'll show how to design two tri-state buffers and connect them to create a tri-state bus. Table 7.3 is the truth table for a tri-state buffer with input, **IN**, output, **OUT**, and tri-state enable, **EN**.

IN	EN	OUT
_	0	High impedance state
A	1	***A***

where _ is a don't care condition and ***A*** is a logic value (0 or 1).

Table 7.3

The circuit we'll design is shown in Figure 7.9. The truth table that defines the tri-state bus of Figure 7.9 is shown in Table 7.4.

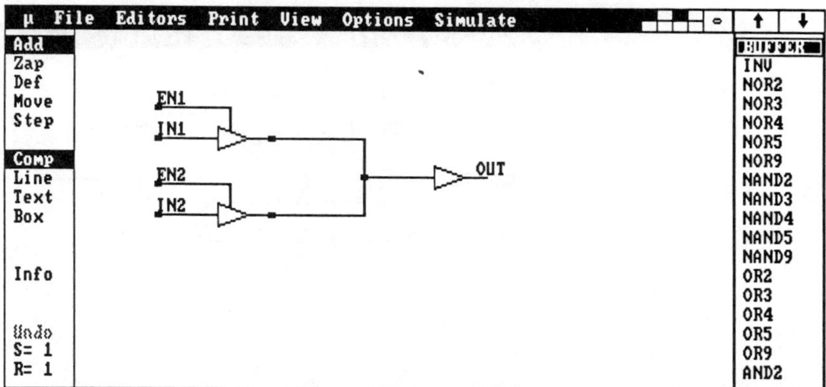

Figure 7.9

EN2	EN1	IN2	IN1	OUT
0	0	_	_	1
0	1	_	A	A
1	0	B	_	B
1	1	_	_	undefined

where _ is a don't care condition and ***A*** and ***B*** are logic values (0 or 1).

Table 7.4

Figure 7.10 shows a circuit that simulates the circuit of Figure 7.9. Notice that an AND gate is used to incorporate the enable signal into the circuit. In addition, extra invertors are required for the circuit to operate properly.

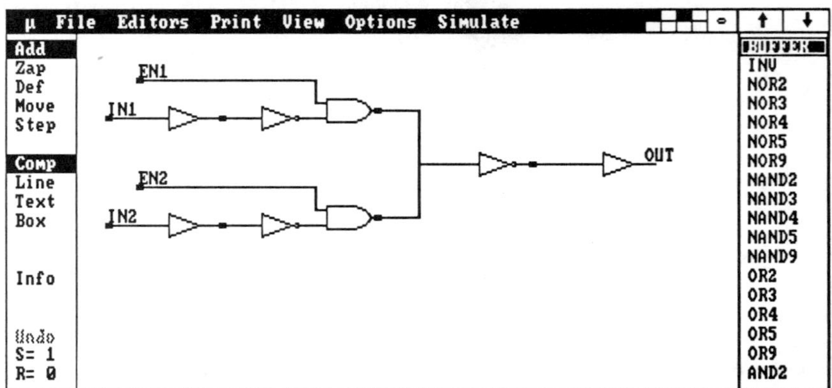

Figure 7.10

Problems

1. Test the correct operation of macro 82, created in this tutorial, by using this macro in a test circuit. You will need five inputs—B2, B1, A2, A1, and C0—and three outputs—S2, S1, and C2—where the most significant bits are B2, A2, and C2.
2. Create a new macro, a four-bit adder, composed of two 82 macros (and additional logic as required). Test the correct operation of the new macro by using it in a test circuit.
3. Create a macro that implements the 74241 TTL package. Use the macro in a circuit to test its correct operation.
4. Test the circuit of Figure 7.10 by drawing it and simulating its operation. Verify that its truth table corresponds to the truth table of Table 7.4.

8 Additional Mouse Features

8 Additional Mouse Features

Tutorial 8 will teach you how to:

- Use the mouse to move components, lines, and text
- Define, move, copy, and zap boxes
- Directly measure waveform timing

The Move Function

The mouse can easily move components, lines, and text in one step, rather than zapping them and then inserting them. Let's use the **BIN_SYNC** file we created earlier to practice using the mouse to move elements. Figure 8.1 shows the **BIN_SYNC** drawing. We'll change the drawing in various ways and not necessarily restore it to its original form. Therefore it is suggested that you not save the drawing at any point in this tutorial.

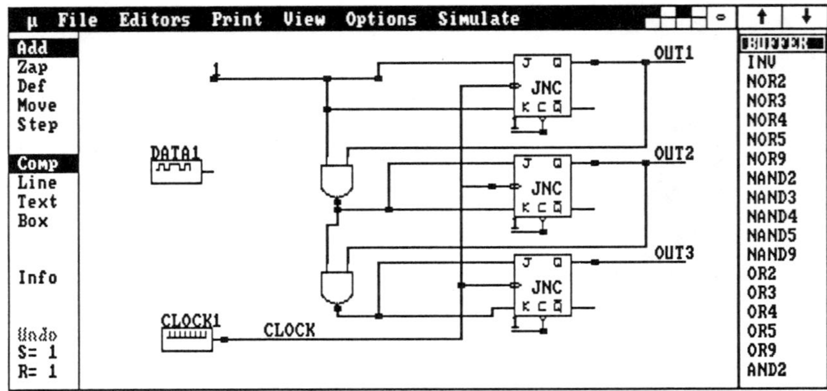

Figure 8.1

Moving Components

Load the drawing **BIN_SYNC** using the File menu. (Remember, you can use the mouse to open a menu and select an option.) To move a component, select **Comp** and **Move** in the Function Selector and move the mouse to the component you wish to move. Try moving the lower AND2 gate. Using the left button, click and drag the component to a new position somewhere to the left. Release the button when the component is properly positioned.

You can scroll the screen while moving a component. That is, if you reach the edge of the Display area while moving a component, the screen will scroll automatically. Try moving the same AND2 gate to the right, completely out of the Display area, by moving the mouse to the gate and then clicking and dragging the component to the right. The screen should scroll as you move the component off the right edge of the Display area.

You can also rotate/reflect a component as you move it. To do this, select the component and click and hold the left button. Clicking the right button will change the rotation/reflection number by one unit per click. Then drag the component to the proper position and release the left button. Try this on the same AND2 component. Notice that the rotation/reflection number changes each time you click the right button.

Now return the AND2 gate to its original position, with a rotation/reflection of 1.

Moving Lines

To move a line, select **Line** and **Move** on the Function Selector. Position the mouse over the line to be moved. In this case try moving the long line under the label CLOCK. Click the left button and drag the line into the desired position. Observe that the way a line moves depends on where the mouse is positioned on the line. Experiment by placing the mouse at different points on the line and then moving the line.

Moving Text

To move text, select **Text** and **Move** in the Function Selector. Position the mouse over the text to be moved. Click and drag the text to its new position. Try moving the label **OUT1** to a different location.

Editing Text

Using the mouse, you can also edit text without moving it. With **Text** and **Def** selected in the Function Selector, position the mouse over the text to be edited and click the left button. A dialog box appears in which the text can be edited by using the four cursor keys, as well as [←], [Del], and [Ins]. Try editing a few of the labels on the screen.

After you finish editing, either press [↵] or click the mouse somewhere in the Display area.

Boxes

A **box** is a rectangular region that can be defined in the Display area using the **Def** and **Box** functions in the Function Selector.

Once a box is defined, its contents can be moved, copied, or zapped. This allows a large number of components to be manipulated in one step. We'll demonstrate some of these functions on the drawing **Box** that came with MICRO-LOGIC II. Load the drawing **Box** using the File menu.

Hint: If at some point in the following steps, you change the diagram irretrievably, simply reload it and start again.

Defining and Adding Boxes

A box is defined by selecting **Box** and **Def** in the Function Selector. The upper left corner of the box is defined by positioning the mouse in the desired position and clicking the left button. The lower right corner of the box is defined by positioning the mouse in the desired position and clicking the right button.

Define a box on the drawing as shown in Figure 8.2 by activating the **Box** and **Def** functions in the Function Selector. Then move the mouse next to the letter **R** and click the right button. Then move the mouse next to the letter **L** and click the left button. Repeat the procedure until your screen looks exactly like Figure 8.2.

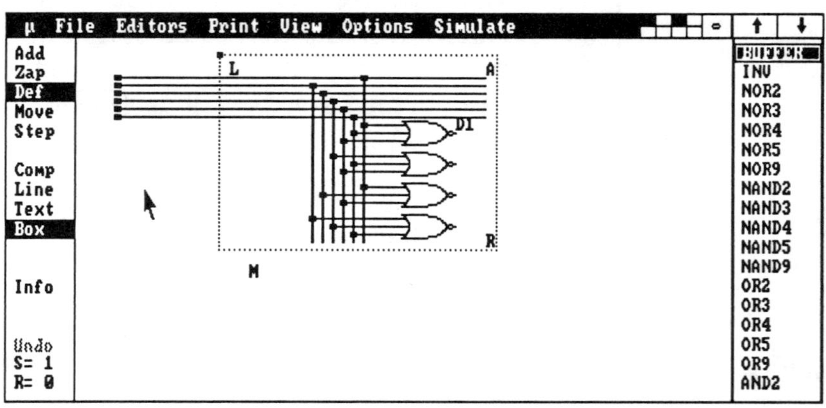

Figure 8.2

Now let's copy the circuitry in the box and place the copy just to the right of the existing box at the letter **A**. Select the **Add** function from the Function Selector. Move the mouse just above the letter **A** and click. In response to the dialog box, click again or press ⏎. (To cancel the Add command, type **N**.)

The copy of the box should now be added as shown in Figure 8.3. If not, select **Undo** in the Function Selector and try again. (You will have to redefine the box.) Notice that once the new box is added, the box definition moves to that box rather than staying with the original box.

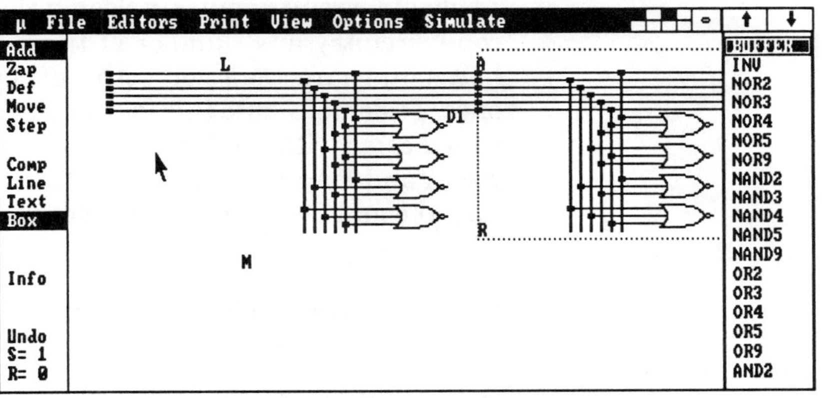

Figure 8.3

Once you've successfully added the box, try zapping the newly added box by selecting **Zap** in the Function Selector. In response to the dialog box, click the mouse again or press ⏎. (To cancel the zap command, type **N**.) The box has now been zapped. If a large dialog box that says **Enter times to step** appears, press Esc.

Boxes can also be moved, which comes in handy if you start to draw a circuit and then find yourself close to the edge of the paper. Define the box as before by selecting the Def function and then locating the upper left and lower right corners. (Note that the letter **R** was deleted when the box was zapped. Position the box in approximately the original position.)

Now move the box by selecting the **Move** function and clicking the mouse in the Display area at the place where you want to position the upper left corner of the box (you may experiment anywhere on the screen). A dialog box appears. (To cancel the Move command, type **N**.) Click the mouse anywhere in the Display area or press ↵ to complete the move. After moving the box somewhere else on the screen, practice moving it back to its original location.

Adding Multiple Boxes

As you've seen, the **Add** and **Box** functions allow you to duplicate circuitry quickly. Using the **Step** function, you can copy circuitry any number of times in either the vertical or horizontal directions, or both vertically and horizontally at once. To demonstrate this feature, let's use the drawing **CELL**, which came with the MICRO-LOGIC II software. Use the File menu and load the drawing **CELL**. As before, if you make a mistake and cannot recover the drawing, simply reload **CELL** and start again.

Let's replicate the circuitry shown on the screen and produce a 3 x 3 matrix of these cells.

Define a box using the **Box** and **Def** functions in the Function Selector such that the box just encloses the circuitry (see Figure 8.4). Now select the **Step** function in the Function Selector.

Figure 8.4

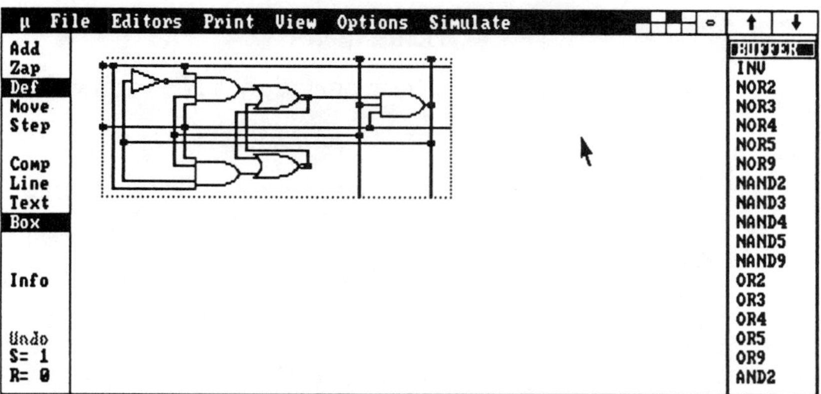

In the dialog box click **Both** so that the **Both** option is selected. Using the keyboard, enter the value **2** in the box located after **Enter times to step**. You have now programmed the Step command to copy the box two times vertically and then two times horizontally. First the box will be redefined around the three vertical circuits and then these three circuits will be replicated two times horizontally. The box will then be redefined around the entire replicated circuit.

Click **Ok** and watch the screen. After the replication is finished you can get a better view of the circuit by selecting option **2:Quadrant** on the View menu. After you finish examining the quadrant, click the quadrant view itself or press ↵.

Measuring Waveform Timing

Now we'll learn a final mouse feature by using the **FULL_ADD** drawing. Load this drawing using the File menu. Open the Simulate menu and select option 1. Set the limits as shown in Table 8.1.

Maximum simulation time in clocks(1-32767)	8
Scale factor	1
Clock period in 1 NS intervals(1-32767)	200
Plot Clock, Intervals, or No grids (C,I,N)	N
Run Asymmetric, Short, Long, or Both delays (A,S,L,B)	S
Run Fixed or Calculated delays (F,C)	F
Name of pattern file	BINARY
Default initial state (0,1,?)	0

Table 8.1

Begin a simulation by clicking the Simulate bubble and, with the fourth and fifth clock cycle on the screen as shown in Figure 8.5, pause the simulation by typing **P**. If you miss these clock cycles, start the simulation over and try again.

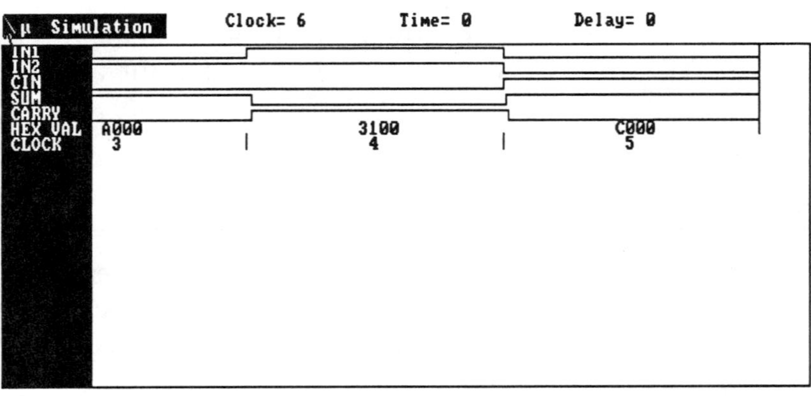

Figure 8.5

Now we'll measure the time between the rising edge of the CIN signal (at the start of the fifth clock cycle) and the falling edge of the CARRY signal just after it. We'll use the mouse to mark these two points on the waveforms. Then, using the **Clock**,

Time, and **Delay** values shown at the top of the screen, we can determine the time difference. Clock should now indicate either 5 or 6 (depending on when you paused the display), and Time and Delay should both be 0.

Move the mouse directly over the rising edge of CIN and click the left button. Notice that a vertical line has appeared, marking the clicked spot. At the top of the screen **Time=** should be set to 0. If the value is not 0 then the mark was not set precisely. (Look at the screen carefully and you'll see that the vertical marker is not exactly over the rising edge of CIN.) If **Clock= 4**, the line must be moved to the right. If **Clock= 5**, the line must be moved to the left. Continue to position and click, using the left button, until **Time= 0**. Notice that the vertical mark and the rising edge of CIN are now perfectly aligned.

Now, using the right button, mark the falling edge of CARRY. Position the mouse directly over the falling edge of CARRY and click the right button. A second vertical mark appears to indicate this position. Make sure the vertical mark is exactly over the falling edge of CARRY. If not, move it by repositioning the mouse and clicking the right button. Once positioned, **Delay= 4** appears. This means that there is a difference of 4 nanoseconds between the rising edge of CIN and the falling edge of CARRY (see Figure 8.6).

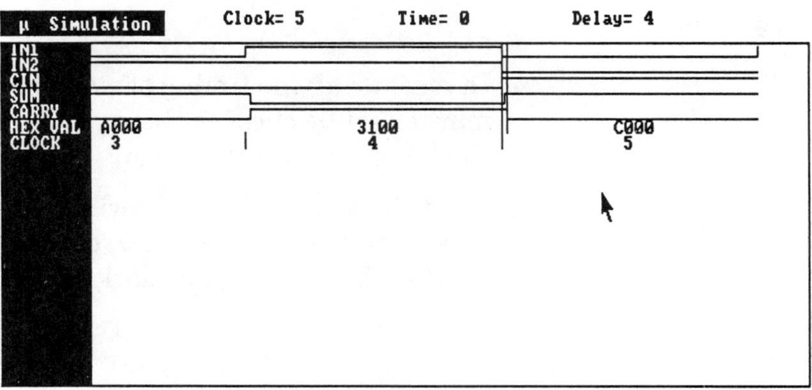

Figure 8.6

The value of **Time** will vary between 0 and (clock cycle -1), in this case 199. **Time** indicates the position within a clock cycle of the marker set by clicking the left button (the specific clock cycle is indicated by **Clock**). **Delay** indicates the absolute difference between the position of the marker set by clicking the left button and the marker set by clicking the right button.

Problems

1. Replicate the drawing **CELL** such that your circuit looks like Figure 8.7.

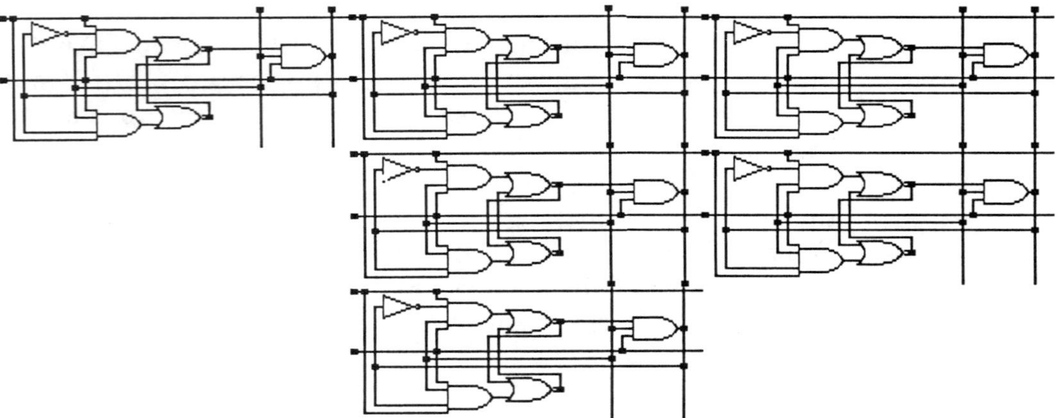

Figure 8.7

2. In measuring the timing of waveforms, does it matter if the marker set by clicking the right button is actually placed to the left of the other marker?

3. Load the **FULL_ADD** drawing and measure the delay between the start of the fourth clock cycle and the rising edge of CARRY in the fourth clock cycle.

Reference Section

1 Quick Reference Guide 165

2 Displays and Pull-Down Menus 173

1 Quick Reference Guide

Starting and Exiting MICRO-LOGIC II

To start MICRO-LOGIC II, type **start** at the MS-DOS prompt. If your computer has a Hercules graphics card, type **start h** instead. If you haven't rebooted the computer since you last used MICRO-LOGIC II, type **start**.

You can exit MICRO-LOGIC II from the μ menu by pressing [Esc]. You can also exit from the Designer Display screen by pressing [Esc].

Filing

Open the File menu by pressing [Alt] **F** or using the mouse. From this menu you can carry out filing functions such as saving or loading a drawing. Type the number of the option you want.

Adding Components

To add a component, select **Comp** and **Add** in the Function Selector by typing **C** and **A**. A dialog box appears. Type the name of the desired component and press ↵.

To add a second component of the same type, press Ins. Ins will add whatever component was the last to be selected.

To add a component with a rotation/reflection value different from the number shown as **R=x** in the Function Selector, type the number before typing **A** or pressing Ins.

Mouse The **Add** and **Comp** functions can be activated by moving the mouse pointer over each function and clicking. They are automatically activated, however, whenever you select a component from the Component Selector by clicking the component name. To add a component to a drawing, click the desired component in the Component Selector. Then move the mouse pointer to the desired position in the display area and click, or click and drag.

To move a component, select **Comp** and **Move** in the Function Selector and click and drag the component to a new position. Release the button when the component is properly positioned. To rotate/reflect a component as you move it, select the component and click and hold the left button. Clicking the right button will change the rotation/reflection number one unit per click.

Deleting Components and Undoing Mistakes

To **Zap** or delete a specific component, select the **Comp** and **Zap** functions on the Function Selector, position the cursor on the component, and type **Z**. Or select **Comp**, position the cursor on the component, and press Del.

You can **Undo** the last command executed by typing **U.**

Mouse Click the **Zap** function in the Function Selector. Click the **Comp** function in the Function Selector. Move the mouse pointer to the component to be deleted and click. To undo a command, click **Undo.**

Adding and Deleting Lines

To add a line, select **Add** and **Line** in the Function Selector. Type **A** to start a new line. Press [Ins] to change direction or to end the line.

To delete a line segment, select **Line** in the Function Selector. Position the cursor on the line and type **Z** or press [Del].

Mouse To draw a line, select **Add** and **Line** in the Function Selector. Position the mouse pointer where the line should start and click the left button. To change direction or to terminate a line, click the right button.

To delete a line segment, select **Zap** and **Line**. Position the mouse pointer on the line and click the left button.

To move a line, select **Move** and **Line**. Position the mouse over the line. Click the left button and drag the line into the desired position.

Adding and Deleting Text

To add text, place the cursor where you want the text to start. Select the **Text** and **Add** functions in the Function Selector. (Instead of typing **T** and then **A**, you can type **T** and then press [Ins].) A dialog box will then ask you for the text to be added. Type the text and press [↵].

To label a line for simulation, place the cursor exactly on the line when you insert the text.

To delete a label, position the cursor on the text and type **Z** or press [Del].

Mouse To add text, select **Add** and **Text** in the Function Selector. Position the mouse pointer where the text is to be placed and click the left button. Type the desired text in the dialog box and press ⏎ or click the mouse anywhere in the Display area.

To delete text, select **Zap** and **Text**. Position the mouse pointer over the text to be deleted and click the left button.

To move text, select **Move** and **Text**. Click and drag the text to its new position.

To edit text without moving it, select **Def** and **Text**. Position the mouse over the text to be edited and click the left button. A dialog box appears in which the text can be edited by using the four cursor keys, as well as ←, Del, and Ins. After you finish editing, either press ⏎ or click the mouse somewhere in the Display area.

Moving and Copying Parts of a Circuit

This can be done with a mouse by defining a box around the region.

Mouse To define a box, select **Box** and **Def** in the Function Selector. The upper left corner of the box is defined by positioning the mouse in the desired position and clicking the left button. The lower right corner of the box is defined by positioning the mouse in the desired position and clicking the right button.

To copy the circuitry in the box, select **Add**. Move the mouse where you want to place the upper left corner of the box and click. In response to the dialog box, click again or press ⏎.

To zap a box, select **Zap**.

To move a box, define the box, select **Move,** and click the mouse where you want to position the upper left corner of the box. Click again or press ⏎.

The **Step** function can replicate circuitry. Define the box and select **Step**. In the dialog box click **Horizontal**, **Vertical**, or **Both**. Using the keyboard, enter the number of times to step. Click **Ok**.

Moving the Cursor and the Drawing Paper

The Drawing View controls at the top right corner of the screen show the current page and quadrant. With [ScrollLock] off, the four cursor keys move the cursor across the page; [PgUp], [PgDn] [Home], and [End] move the display area up, down, left, and right, respectively. With [ScrollLock] on, the cursor keys scroll the paper; [PgUp], [PgDn], [Home], and [End] move the display area diagonally.

The **S= x** near the bottom of the Function Selector is the grid step number indicator. Pressing [+] doubles the value of S; pressing [−] resets S to 1.

> **Mouse** To reset the grid step number to 1, position the mouse pointer over **S= x** and click.

Component Definition

To examine component characteristics, activate the **Info** function on the Function Selector, move the cursor to a component, and type **I**. The library information of that component will be displayed.

To search for either a component or a specific piece of text in a drawing, use option **7:Search** in the View menu.

To change the definition of existing components or to create new components (or macros), access the Component editor by selecting **1:Components** on the Editors menu.

Simulation

The Pattern Editor and Data Channel Definition

To reach the Pattern editor, choose **3:Patterns** on the Editors menu. The states of a data channel can be defined in two ways:

- Direct keyboard editing: Use the cursor keys (or the mouse) to reach the digits you want to change. Then enter the data directly using the keyboard.
- The Patterns menu: Choose Random pattern, Binary pattern, Invert pattern, or Set a block by typing the first letter.

Type the number of the data channel you want to change and press [↵]. If you make a mistake, press [Esc] to cancel the selection.

Reports

On the Print menu, select **2:Print reports**. There are three kinds of reports:

- Netlist report: lists all components used in the drawing and their network connections.
- Unconnected pins report: lists any pins that are not connected.
- Capacity report: lists the usage of each type of component. This is also included in the netlist report.

Simulating a Circuit

Display the circuit you want to simulate. In the Simulate menu, choose option **1:Simulate.** The Analysis Limits screen will be displayed. Edit the values as necessary.

To choose which nodes to monitor, open the Simulate menu and select **E:Edit nodes monitored**. The Nodes Monitored screen will be displayed. List each node on a separate line. To leave this screen, press [Esc] or open the Simulate menu. (Option **A:Edit analysis values** will return you to the Analysis Limits screen.)

To begin a simulation, press [Home] or select **Home: Simulate** on the Simulate menu. The waveforms will be displayed.

To redefine the clocks, select **C:Clocks** on the Simulate menu.

To set patterns, select **P:Patterns** on the Simulate menu.

Measuring Waveform Timing

This can be done with the mouse while a simulation is running.

Mouse Begin a simulation. When the clock cycles you wish to measure appear on the screen, pause the simulation by typing **P**. If you miss these clock cycles, begin the simulation over and try again. Mark the first point by clicking the left button. Mark the second point by clicking the right button. Read the **Delay** in nanoseconds.

2 Displays and Pull-Down Menus

Designer Display Screen

The Designer Display screen (see Figure R.1) lets you create drawings, which can then be saved, retrieved, edited, refiled, deleted, and so on. All functions of MICRO-LOGIC II are accessed from the Designer Display screen, including print, plot, and simulation options.

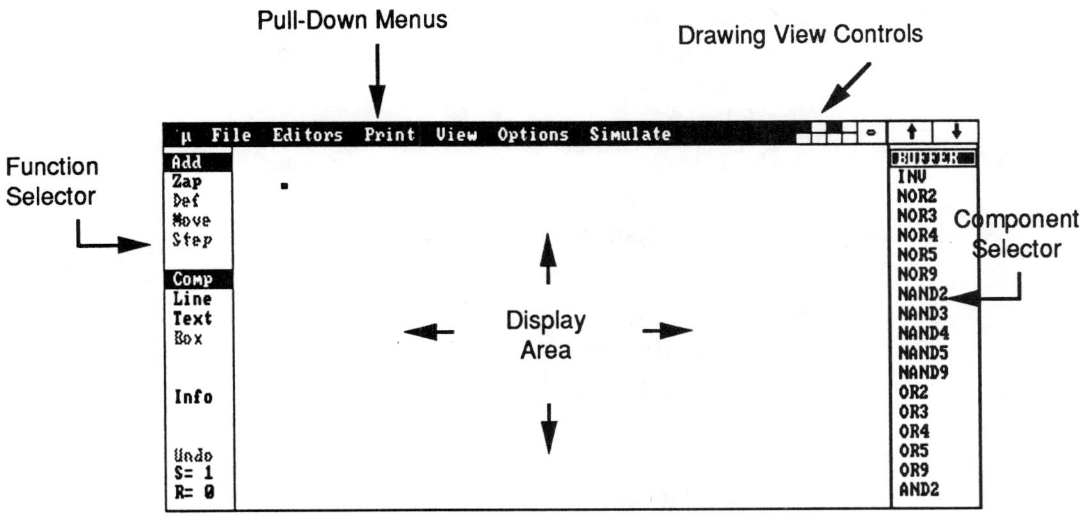

Figure R.1

The Designer Display screen is divided into five major areas:

- Pull-down menus
- Function Selector
- Component Selector
- Drawing View controls
- Display area

Pull-Down Menus

Using the pull-down menus on the Design Display screen, you can select the following major functions:

- μ (See Figure 1.2.)
- File menu
  - Save drawing
  - Load drawing
  - Erase drawing
  - Merge drawing
  - New drawing
  - Change data drive
  - Update macros
  - File warning
  - File dialog box
- Editors menu
  - Components
  - Shapes (not available in student version)
  - Patterns
  - Clocks
- Print menu
  - Print drawing
  - Print reports
  - Plot drawing (not available in student version)
- View menu
  - Capacity used
  - Quadrant
  - Quadrants
  - Show component text
  - Show grid text

- Show component numbers
- Search
- Options menu
  - Wired-NOR
  - Wired-OR
  - Colors (not available in student version)
  - Help lines
  - Mouse ratio
- Simulator module
  - Simulate

Function Selector

Use this window to select an action and element for subsequently constructing a diagram.

Actions include:

- Add
- Zap
- Def (mouse only)
- Move (mouse only)
- Step (mouse only)

Elements include:

- Comp
- Line
- Text
- Box (mouse only)

Functions include:

- Info
- Undo
- S=:
- R=:

Component Selector

Use this window to select a component from the Component library for subsequent diagram construction. You can also save, delete, or modify components in the Component library.

Drawing View Controls

Use this window to identify the current page and quadrant. Also use this window to select specific pages and quadrants for viewing and to scroll the drawing window in a number of directions (mouse only).

Display Area

Use this area as the drafting paper on which your logic diagram is drawn.

Component Editor

The Component editor (see Figure R.2) manages the Component library and can create almost any component. Use the Component library to retrieve, modify, or add to library content.

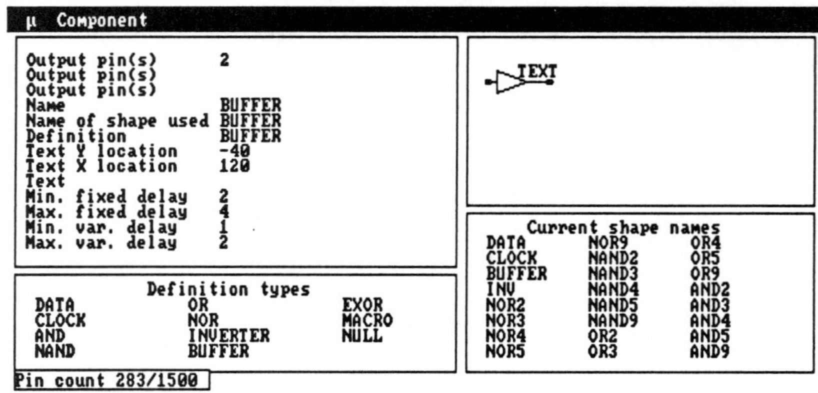

Figure R.2

The Component editor display is divided into five major areas:

- Pull-down menu
- Component display
- Definition types
- Shape name list
- Edit window

Pull-Down Menu

The Component editor pull-down menu lets you select among the following major functions:

- Select
- Find
- Hardcopy
- Move component

Component Display

Use this window to show the shape selected and the specified text field of the chosen component.

Definition Types

Use this window to list acceptable entries for the definition field of the component.

Shape Name List

Use this display to select a shape from the Shape library list for the component.

Edit Window

Use this window to modify important information about the component. The fields available for editing are as follows:

- Output pin(s)
- Name
- Shape name
- Definition
- Text location
- Text
- Min. fixed delay (Boolean components only)
- Max. fixed delay (Boolean components only)
- Min. variable delay (Boolean components only)
- Max. variable delay (Boolean components only)
- Clock number (clocks only)
- Data number (data channels only)

Shape Editor

The Shape editor creates and maintains the shapes used to draw components. It is not available in the student version of MICRO-LOGIC II.

Pattern Editor

The Pattern editor creates data patterns for data channels, which drive networks during a simulation. Patterns are stored in files rather than in a library; they are always 32 channels wide and 1024 clock cycles long.

The timing relationships between data channel waveforms and clock waveforms are shown in Figure R.3.

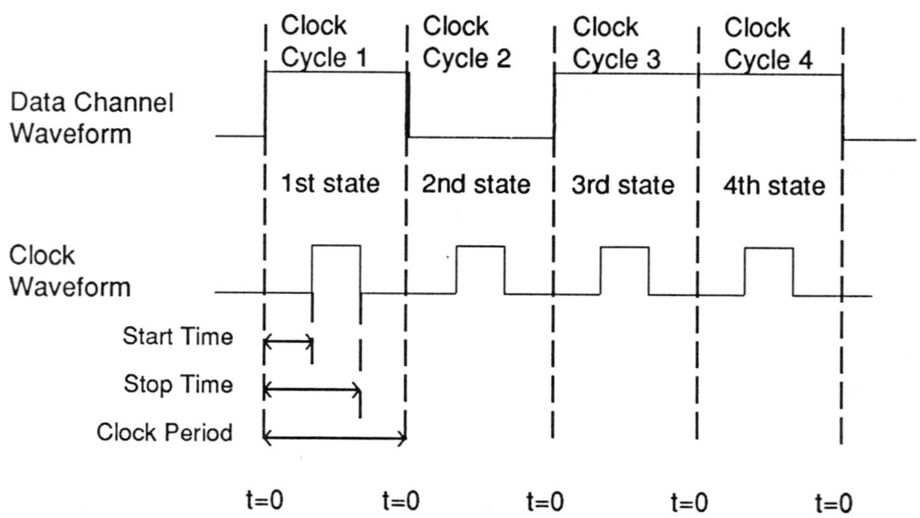

Figure R.3

The Pattern editor display (see Figure R.4) contains two major areas:

- Pull-down menu
- Pattern window

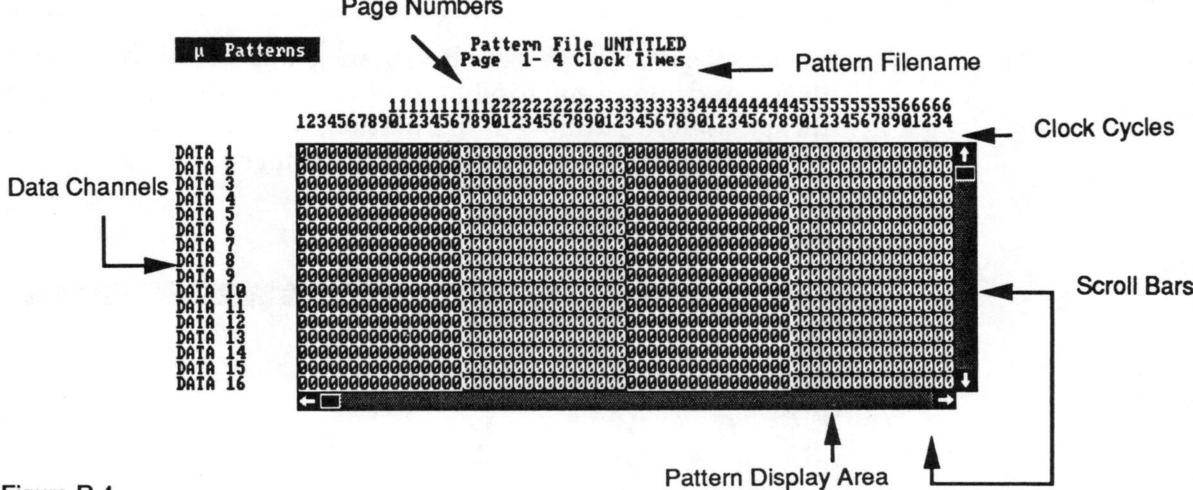

Figure R.4

Pull-Down Menu

The Pattern editor pull-down menu contains the following major functions:

- μ
- Patterns

Selecting Patterns allows you to access the following options:

- Save pattern
- Load pattern
- Erase pattern
- Hardcopy
- Random
- Binary
- Set
- Invert

Pattern Window

The Pattern window displays the clock cycles and waveforms on the screen, scrolling the information as it is generated.

Clock Editor

The Clock editor (see Figure R.5) manages the Clock Waveform library and allows you to edit any of the clock waveforms in the library. Before running a simulation, you must specify the Start time interval and the Stop time interval.

```
 µ                    Clock waveform editor
Name              Start time   Stop time
Clock1            10           30
Clock2            10           30
Clock3            10           30
Clock4            10           30
Clock5            10           30
Clock6            10           30
Clock7            10           30
Clock8            10           30
Clock9            10           30
Clock10           10           30




Esc:Exit editor Tab:change columns
```

Figure R.5

The Clock editor is divided into two major areas:

- Pull-down menu
- Edit fields

Pull-Down Menu

The Clock editor pull-down menu consists of the standard µ menu.

Edit Fields

The Edit fields appear below the pull-down menu. From here you can edit, create, or delete the clocks you select.

Simulator Module

The Simulator module (see Figure R.6) lets you perform a timing analysis on the logic drawings that you create.

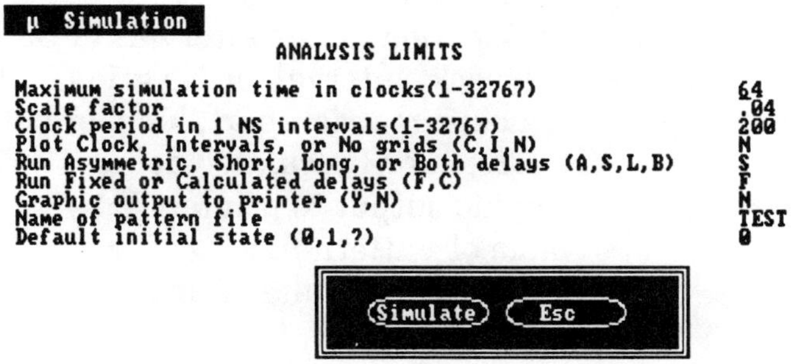

```
µ Simulation
                    ANALYSIS LIMITS
Maximum simulation time in clocks(1-32767)                    64
Scale factor                                                  .04
Clock period in 1 NS intervals(1-32767)                       200
Plot Clock, Intervals, or No grids (C,I,N)                    N
Run Asymmetric, Short, Long, or Both delays (A,S,L,B)         S
Run Fixed or Calculated delays (F,C)                          F
Graphic output to printer (Y,N)                               N
Name of pattern file                                          TEST
Default initial state (0,1,?)                                 0

                    (Simulate) (Esc)
```

Figure R.6

The Simulator display is composed of two major areas:

- Pull-down menu
- Analysis list

Pull-Down Menu

The Simulator pull-down menu lets you select between the following major functions:

- µ
- Simulations

If you select Simulate, you can access the following options:

- Simulate:Start run
- Edit analysis values
- Edit nodes monitored
- Patterns
- Clocks
- Save run (Not available in the student version.)
- Retrieve run (Not available in the student version.)
- Text file

Analysis List

Use the Analysis list to modify the analysis limits of a logic drawing. Analysis limits control the simulation run.

If you access Analysis list, the following analysis limits are displayed:

- Maximum simulation time in clocks
- Scale factor
- Clock period in 1ns intervals (1-32767)
- Plot clock, Interval, or No grids (C,I,N)
- Run asymmetric; short, long, or both delay (A,S,L,B)
- Run fixed or Calculated delays (F,C)
- Graphic output to printer (Y,N)
- Name of Pattern file
- Default initial state of unconnected pins (0,1,?)

3 Error Messages

Designer Display Screen Error Messages

Error...box regions exceed boundaries A Box operation was attempted that would have resulted in some drawing elements being placed outside the maximum permissible boundaries. This error usually occurs during an Add box operation when the point of insertion is such that a portion of the box being copied would go outside the boundaries.

Error...can't write file *name* The requested file save operation cannot be performed because of one of the following conditions:

- The filename is bad; use only alphanumeric characters for filenames.
- The disk drive door isn't closed.
- The data disk is full.

Error...coordinates out of range The user attempted to merge a file when the elements of the file are being placed into the drawing with an offset that places them out of the current quadrant.

Error...can't read file *name* An attempt to read a file fails for one of the following reasons:

- The filename is bad; use only alphanumeric characters for filenames.
- The disk drive door isn't closed.
- The file doesn't exist or is unreadable.
- The file is a macro netlist file and has never been generated by saving the drawing version of the file.

Error...too many components The user entered more than the maximum number of components.

Error...too many labels The user entered more text labels than the system can hold.

Error...too many lines The user entered more than the maximum number of line segments.

Error...too many nodes Netlist generation produces more than the maximum permissible number of nodes.

Component Editor Error Messages

Error...can't write the Component library file The system unsuccessfully attempted to file the Component library after edits were made. This error usually occurs only when a disk is full or when the drive door has been left open.

Pattern and Clocks Editor Error Messages

Error...can't save the Clock library file The system attempts to file the Clock library after edits are made. This error usually occurs only when the disk is full or when floppy drives are being used and the drive door has been left open.

Error...can't write the file *name* The system attempts to save a pattern file. This error usually occurs only when the disk is full, when a bad filename is supplied by the user, or when floppy drives are being used and the drive door has been left open.

Error...can't read the file *name* The system attempts to read a pattern file. This error usually occurs only when a file is corrupted or when floppy drives are being used and the drive door has been left open.

Simulator Error Messages

Error...can't read file *name* The system can't read a pattern file. The user should check the spelling and be sure that no extensions are included.

Error...can't save file *name* The system can't save a file, probably because there isn't enough space on the disk.

Error...fanout capacity exceeded The fanout portion of the netlist exceeds the maximum. The only solution is to eliminate some portion of the drawing and thus reduce the total fanout list.

Error...maximum number of events exceeded This error occurs when there are too many unprocessed events. As the system generates events, they are placed in an event queue. As they are evaluated at the current simulation time, they are removed. If, at a particular point in the simulation, more events are generated than are removed, the queue can become full. If this occurs, the user must remove some of the drawing, change the data channels, or reduce the number of clocks to reduce the peak usage of the event queue.

Error...maximum simultaneous events exceeded Too many simultaneous events are present. This error can occur if one component, usually a clock or data channel, is driving a great many other components. Since the system can handle as many as 1200 simultaneous events, this problem occurs frequently. When it does, separate the common line with a low delay buffer to push the corresponding events out to a slightly later time slot.

Error...too many clocks The drawing contains too many clocks. Ten different clocks are available. However, a particular clock can be used more than once. The maximum number of clocks used in a drawing should not exceed 20.

Error...too many data channels The drawing contains too many data channels. Thirty-two different data channels are available; a particular data channel can be used more than once. The maximum number used in a drawing should not exceed 64.

Range error in Maximum simulation time The designated value either exceeds 32767 or is less than 1.

Range error in Scale factor The designated value is less than or equal to zero.

Range error in Clock period The designated value either exceeds 32767 or is less than the largest Stop time of clocks in the network.

Range error in Delay options (A,S,L,B) The designated value is not A, S, L, or B.

Range error in Printer option (Y,N) The designated value is not Y or N.

Range error in Delay options (F,C) The designated value is not F or C.

Range error in Pattern file *name* The designated value is not a valid filename.

Range error in Initial state options (0,1,?) The designated value is not 0, 1, or ?.

Warning...text label has a zero node The text label of a node to be monitored has a zero value, indicating that the desired text label used for the node is off the node line or gate output of the node. Check the drawing to ensure that the label is on the line.

4 Components in the Library

BUFFER	
INV	Invertor
NOR2	
NOR3	
NOR4	
NOR5	
NOR9	
NAND2	
NAND3	
NAND4	
NAND5	
NAND9	
OR2	
OR3	
OR4	
OR5	
OR9	
AND2	
AND3	

AND4	
AND5	
AND9	
EXOR	
LATCH	Latch type FF
DN	D FF, Negative edge triggering
DNC	D FF, Negative edge triggering, and Clear
DPC	D FF, Positive edge triggering, and Clear
DPPC	D FF, Positive edge triggering, Preset, and Clear
JMC	JK FF, Master-slave triggering, and Clear
JMPC	JK FF, Master-slave triggering, Preset, and Clear
JNC	JK FF, Negative edge triggering, and Clear
JNP	JK FF, Negative edge triggering, and Preset
JNPC	JK FF, Negative edge triggering, Preset, and Clear
RM	RS FF, Master-slave triggering
RMC	RS FF, Master-slave triggering, and Clear
RMPC	RS FF, Master-slave triggering, Preset, and Clear
TMC	T FF, Master-slave triggering, and Clear
TMPC	T FF, Master-slave triggering, Preset, and Clear
42	BCD-decimal decoder
74	Dual D FF PE triggered with Preset and Clear
86	Quad 2-input exclusive-or gates
90	Decade counters, divide by 2 and by 5
112	Dual JK NE triggered with Preset and Clear
138	3-to-8 line decoders
139	Dual 2-to-4 decoders
145	BCD-to-decimal decoder
151	1 of 8 data selectors/multiplexers
153	Dual 4-line to 1-line data selectors
154	4-line to 16-line decoders/demultiplexers
157	Quad 2 to 1 line data selectors/multiplexers

160	Synchronous 4-bit decade counters
175	Quad D flip-flops
190	Synchronous up/down BCD counters
194	4-bit bidirectional universal shift registers
257	Quad data selector/multiplexers
283	4-bit binary full adders
5NS	
15NS	
20NS	
DATA1	
DATA2	
DATA3	
DATA4	
DATA5	
DATA6	
DATA7	
DATA8	
DATA9	
DATA10	
DATA11	
DATA12	
DATA13	
DATA14	
DATA15	
DATA16	
CLOCK1	
CLOCK2	
CLOCK3	
CLOCK4	
CLOCK5	

5 Files on Disk

The following files are included on the program disk:

README.DOC	Latest program information
START.BAT	Startup batch file
ML2.EXE	Startup program
BUFFER.LOG	User system options file
INT10.COM	Hercules graphics interface
BRUN30.EXE	Runtime file
RESSTRPT.LOG	System data files
RESOURCE LOG	System data files

The following files are included on the data disk:

CLOCK.LOG	User Clock library
COMLIBB.LOG	User Component library
COMLIBA.LOG	User Component library
BITMAPCS.LOG	Shape library, Bitmaps
BITMAPCL.LOG	Shape library, Bitmaps
BITMAPAS.LOG	Shape library, Bitmaps

BITMAPAL.LOG	Shape library, Bitmaps
SHAPELST.LOG	Shape library, Primitives list
SHAPETAG.LOG	Shape library, Text list
SHAPEPIN.LOG	Shape library, Pin list
EXAMPLE.DWG	Sample drawing
BOX.DWG	Sample drawing
CELL.DWG	Sample drawing
MACTEST.DWG	Sample drawing
DNC.DWG	Sample drawing
DNC.MAC	Sample macro
DNCTEST.DWG	Sample drawing
SIMEX.DWG	Sample drawing
WIRE_OR.DWG	Sample drawing
TEST.PAT	Sample pattern file
BINARY.PAT	Sample pattern file
GATES.TMP	Compressed macro drawings
LINEH.TMP	Compressed macro drawings
LINEV.TMP	Compressed macro drawings
STRINGS.TMP	Compressed macro drawings
KEY.TMP	Compressed macro drawings
UNPACK.EXE	Unpacker for creating macro drawing files

When copying the data disk to either a floppy disk or a hard disk, the following files should not be copied, or should be erased from the copy. See Part I, Getting Started, for directions.

GATES.TMP
LINEH.TMP
LINEV.TMP
STRINGS.TMP
KEY.TMP
UNPACK.EXE

UNPACK should be used to unpack the compressed macro drawing files onto the data disk, but the compressed files themselves and the unpacker should not be copied onto the data disk.

Once you complete the tutorials, you may wish to delete some of the files from your copy of the data disk in order to provide more room for your own data files. Directions for erasing files appear in Part I, Getting Started. Only the following files must be included on the data disk for the system to operate correctly:

CLOCK.LOG
COMLIBB.LOG
COMLIBA.LOG
BITMAPCS.LOG
BITMAPCL.LOG
BITMAPAS.LOG
BITMAPAL.LOG
SHAPELST.LOG
SHAPETAG.LOG
SHAPEPIN.LOG

Glossary

Note: Features available only with the mouse are enclosed in brackets [ ].

Add
Action located in the Function Selector window. Adds a component, line, piece of grid text, [or circuitry box to the existing drawing. Use this command to duplicate repetitive circuitry. When duplicating boxes, the system copies the entire contents of the box, except text, to the location where the mouse is clicked.]

Analysis limits
The factors that the user defines and that control the simulation run.

Binary
Option of the Patterns menu of the Pattern editor, used to create a binary pattern in the specified data channel position using a specified binary divider, which is equal to the distance in clock cycles between successive 1 (one) states.

[Box]
Element located in the Function Selector window that refers to all elements inside the boundaries of the currently defined box.

Capacity report
Option of the Print Reports function, used to list the usage of each type of component, the node, and the fanin (simulation capacity used).

Capacity used

Function of the View menu of the Designer Display screen. Tells the user the amount of drawing capacity used by the displayed drawing.

Change drive

Function of the File menu of the Designer Display screen. Allows the user to change the disk drive; the new designated drive must contain all essential library files. These files are listed in Chapter 4 of the Reference Section.

Clock editor

The module that manages the Clock Waveform library. Clock waveforms are available in as many as 32767 intervals. A clock waveform must be chosen before a simulation is run.

Clock period in 1-nanosecond intervals (1-32767)

A user-specified number of time intervals with a nominal value of 1 nanosecond each that determines the period of the clock cycle in real time.

Clock waveform

The time limits specified during a clock cycle, based on the Start time interval and Stop time interval designated by the user. Each interval is nominally assigned 1 nanosecond. The user specifies the number of 1-nanosecond intervals per clock cycle at the time of simulation; this specifies the clock cycle in real time units. Clock waveforms are specified during a clock cycle as follows: 0 from t = 0 to t = Start time; 1 from t = Start time to t = Stop time; t = 0 from t = Stop interval to t = Clock cycle length.

Clocks

Option of the Simulation submenu of the Simulator pull-down menu and an option of the Editor menu of the Designer Display screen used to access the Clock editor. The Clock editor contains the library of clock waveforms specified for a simulation. Clock patterns must be stored in named files within the library and specified in the analysis limits before a simulation can be run.

Colors

Function of the Options menu of the Designer Display screen. Displays the color palette of the user's hardware and lets the user select background and foreground colors; choices may be saved as options. For EGA-equipped systems only. (EGA capability is not available in the student version of MICRO-LOGIC II.)

Comp Element located in the Function Selector window that refers to any component currently defined in the Component library, including data channels, clocks, Boolean functions, and macros.

Comp numbers Function of the View menu of the Designer Display screen. Displays the number associated with a component.

Comp text Function of the View menu of the Designer Display screen. Displays the text associated with a component.

Component editor Function of the Editors menu of the Designer Display screen. An editing system that both manages the Component library and creates almost any component. The Component library contains a large set of predefined components that can be accessed without having to construct them; components may also be constructed element by element.

Component Selector Area located at the far right of the Designer Display screen; used to review [and/or select] the current component.

Component window Area at the upper right of the Component editor display screen. Used to show the component shape selected and the location of the specified text field; shown to drawing scale. Squares denote input pins and circles denote output pins.

Components Elements found in the Component library; used to construct drawings of circuits, including data channels, clocks, Boolean gates, and macros.

Data channel A source of 1024 states fed into the node(s) to which it is attached. Data channels must be specified before a simulation is run.

[Def] Action located in the Function Selector window. Defines the boundary of the current box. Used to change a component from one type to another or to change existing grid text.

Default initial state of unconnected pins (0,1,?)
Analysis field that assigns a value to unconnected pins of gates. If the user chooses ?, the system will produce uncertain intervals in the output waveform of any gate with unconnected pins.

Definition types display
Area to the bottom left of the Component editor display screen; used to list the set of acceptable entries for the Definition field.

Designer Display screen
Display screen that allows the user to create drawings, using either mouse-based or keyboard commands. All other functions of the system are accessed through the Designer Display screen, including print, file, and simulation options. The Designer Display screen has five major areas: pull-down menus; Function Selector; Component Selector; Drawing View controls; and Display area.

DIP
Dual In-line Package shape, used in defining macros.

Disk
Option of Print Reports function, used to send the report to a disk file called **NAME.NET**, where NAME is the name of the displayed drawing.

Drawing View controls
Area of controls located at the upper right corner of the Designer Display screen. The view controls indicate the current page and quadrant [and allow the user to select the quadrant and page at which to view the drawing].

Edit analysis values
Option of the Simulation submenu of the Simulator pull-down menu; used to activate the Analysis Limits screen.

Edit nodes monitored
Option of the Simulation submenu of the Simulator pull-down menu; used to edit the list that defines the nodes monitored during simulation.

Edit window
Area in the Component editor; contains all the important information about components and can be edited. The following fields are available for editing in the Edit window: Output pin(s), Name, Shape name, Definition, Text Y location, Text X location, Text, Min. fixed delay, Max. fixed delay, Min. variable delay, and Max. variable delay (delay parameters are used exclusively by Boolean functions).

Editors Menu located in the Designer Display screen; used to access the following edit functions: Components, Shapes, Patterns, and Clocks.

Erase drawing Function of the File menu in the Designer Display screen. Erases a drawing from the disk; the user selects the file from the File dialog box.

Erase pattern Option of the Patterns menu of the Pattern editor; used to erase a Pattern file from disk through the File dialog box.

Esc Key used to exit menus, editors, the Simulator, and MICRO-LOGIC II.

Expand Option of the Runtime menu of the Simulator Display screen; used to double the simulation scale factor, thereby stretching the waveforms to make delays more apparent.

Expand macros Option of the Print Reports function; used to expand the contents of any macros in the drawing; reports only data channels, clocks, and Boolean gates. The system will list components in the drawing only if this option is not selected.

File Menu located in the Designer Display screen; used to perform the following file options: Save drawing, Load drawing, Erase drawing, Merge drawing, New drawing, Change drive, Update macros, File warning, File dialog box.

File dialog box Function of the File menu of the Designer Display screen. Enables/disables the File dialog box, which contains a listing of all available drawing files. If disabled, the system will not display a listing, but rather request the desired filename to be loaded.

File warning Function of the File menu of the Designer Display screen. Enables/disables file warning, which occurs when an altered file has not been saved.

Find Option of the Component editor pull-down menu; used to search for and display the selected shape name. Not available in the student version of MICRO-LOGIC II.

Flag

Function that the user can perform to scroll through library files quickly. [The user marks certain components that are used frequently during drawing, allowing the system to scroll and display only the names of the components marked by the user. To flag a component, click the right mouse button over the component name, which will outline the name with a box. Any component name so outlined is flagged for quick reference.] The user accesses this subset by pressing ⊟, which allows the system to scroll through the flagged subset only. Flagged components are saved when the user quits the program.

Function Selector

Area located at the left of the Designer Display screen; used to select the action and element of subsequent mouse/keyboard movements and clicks. The user must select an action to perform and an element on which to perform it in conjunction with constructing circuits.

Graphic output to printer (Y,N)

Analysis limit option that lets the user choose whether to produce a graphics dump of the waveforms monitored; Epson-compatible printers only.

Grid step number

The increment at which scrolling occurs; the current step amount is shown near the bottom of the Function Selector window (S=:). Pressing ⊟ on the numeric keypad resets the scroll step to 1 grid; pressing ⊞ on the numeric keypad doubles the previous scroll step set.

Grid text

Function of the View menu of the Designer Display screen. Displays the text associated with a grid to the user.

Hardcopy

(1) Option of the Patterns submenu of the Pattern editor pull-down menu; used to create a printed copy of the entire pattern currently displayed. (2) Option of the Component editor pull-down menu; used to produce a printed list of library components.

Help lines

Function of the Options menu of the Designer Display screen. Offers one-line descriptions of the activated function.

Home:Start run

Option of the Simulation submenu of the Simulator pull-down menu, used to initiate a simulation run by clicking the mouse or pressing [Home].

Info

Function located in the Function Selector window. Provides information on any component chosen by the user through a mouse click/keyboard selection.

Initial States Evaluator

System monitor that evaluates the initial states of a drawing prior to running a simulation. Sets each clock and data channel to its initial state, sets all unconnected pins to the specified default state, and sets any proscribed nodes to the specified state. When all states are stable, the system can then run the simulation. If the system cannot stabilize all initial states, then the user can press [Esc] to exit the evaluator routine and enter the simulation engine routine.

Invert

Option of the Patterns menu of the Pattern editor; used to invert the current pattern at a specified data channel position.

Line

Element located in the Function Selector. Refers to any line segment in a drawing.

Load drawing

Function of the File menu in the Designer Display screen. Loads a drawing from disk. [The user can flag a filename for faster selection by clicking the right mouse button when the filename is highlighted; this places the file in a small subset accessed when [⇆] is pressed.]

Load pattern

Option of the Patterns menu of the Pattern editor; used to load a pattern file from disk from the File dialog box.

Macro

A drawing with text labels that defines the connections with the pins of the shape that represents the macro drawing. The text labels **PIN1, PIN2, ... PIN40**, reserved for this purpose, are referred to as **PINm** text labels.

Max fixed delay

The maximum fixed delay of the gate; if the asymmetric simulation delay option is chosen, this value represents the fixed delay associated with a 0 to 1 transition.

Max variable delay — The maximum variable delay parameter, used to compute fanout-dependent delays for the maximum case; if the asymmetric simulation delay option is chosen, this value represents the fanout-dependent delay associated with a 0 to 1 transition.

Maximum simulation time in clocks — The duration of a simulation specified in terms of the number of clock cycles to run. In addition to the pattern files of 1024 clocks, the user may add more clocks by using files designated **NAME1, NAME2**, and so on for the initial pattern file **NAME**—as many as 31 times, until the maximum simulation time of 32767 is reached.

Merge drawing — Function of the File menu in the Designer Display screen. Merges a drawing on the disk with the one currently on the display screen, using the cursor or mouse location as the origin.

Min fixed delay — The minimum fixed delay of the gate; if the asymmetric simulation delay option is chosen, this value represents the fixed delay associated with a 1 to 0 transition.

Min variable delay — The minimum variable delay parameter, used to compute fanout-dependent delays for the minimum case; if the asymmetric simulation delay option is chosen, this value represents the fanout-dependent delay associated with a 1 to 0 transition.

[Mouse ratio] — Function of the Options menu of the Designer Display screen. Affects how far the mouse arrow moves in response to the user's actions; offers a high and low value for the mouse ratio.

[Move] — Action located in the Function Selector. Allows the user to move elements within the drawing. Perform this function using a click and drag action. When moving a line, initiate action between two corners of the line. When moving grid text, initiate action on the text itself.

Move component — Option of the Component editor pull-down menu; used to move a component to another place in the list. Affects the placement of the component only in the Editor and Designer Display screen component selectors.

Name — The name of the component, limited to eight alphanumeric characters.

Name of pattern file — The analysis field that contains the name of the pattern file to be used by the data channels present in the network; the file should be on the data disk before the simulation is run.

Netlist — A listing of information for a component. Shows the component's library number, type, and the node numbers of the gates connected to its inputs.

Netlist report — Option of the Print Reports function, used to list all components used in the drawing and their netlists.

New drawing — Function of the File menu in the Designer Display screen. Clears the display screen in preparation for a new drawing.

Nodes monitored screen — Screen accessed from the Simulation submenu of the Simulate menu. Lists the text labels of the nodes whose waveforms are to be monitored during simulation; 19 is the maximum possible number of text labels.

Null — In the Component editor, a component type used as a place marker; not a usable component.

Options — Menu located in the Designer Display screen; used to perform the following functions: wired-NOR, wired-OR, colors, help lines, and mouse ratio.

Origin — The specific point in a component graphics symbol that will be positioned at the cursor [or mouse pointer] when the component is added to a drawing.

Output pin(s) — The fields that specify the set of pins that are to provide outputs for the component. Pin numbers must match the actual outputs. For macros, the set of output pins must match the actual connection of gates specified in the macro drawing.

Pattern editor — The module that saves, edits, and reviews data channel pattern files, used when running simulations. The Pattern editor files contain 1024 states for each of 32 data channels; these files can be linked in series for simulations as long as 32767 time intervals. The Pattern editor can be activated from the Editors menu of the Designer Display screen or the Simulation submenu of the Simulate pull-down menu.

Patterns — Menu of the Pattern editor. Used to provide the following pattern options: Save pattern, Load pattern, Erase pattern, Hardcopy, Random, Binary, Set block, and Invert.

Pause — Option of the Runtime menu of the Simulator Display screen; used to let the system pause at the end of the next clock cycle. [When the system is paused, the user may measure time intervals with the mouse.]

Plot circuit — Function of the Print menu of the Designer Display screen. Not available in the student version of MICRO-LOGIC II.

Plot Clock, Interval, or No Grids (C,I,N) — The analysis limit that controls whether clock cycle and time interval markers will be added to the display.

Plotter — The module that creates the plotted drawings.

Pull-down menu — Located in the topmost area of the screen; used to select major functions within the programs and modules of MICRO-LOGIC II.

Print — Menu located in the Designer Display screen; used to perform the following print options: Print drawing, Print reports, and Plot drawing. Plot drawing not available in the student version of MICRO-LOGIC II.

Print drawing — Function of the Print menu of the Designer Display screen. Provides a variety of graphics dumps of the drawing to the printer.

Print reports — Function of the Print menu of the Designer Display screen. Tells the user the simulation capacity used, the netlist, and the incidence of unconnected pins of the drawing. Specific options within this function are: Netlist report, Unconnected pins report, Capacity report, Expand macros, Track lines, Disk, Screen, and Printer.

Printer
Option of the Print Reports function; used to send the report to the printer only.

Propagation delay
The time difference between a state change on an input pin and the corresponding state change at the output pin; only Boolean gates have delays.

Proscribed node
Any node to which is attached a text label of 0, 1, or ?. Proscribing a node will work only for unconnected input pins; a pin connected to an output will be assumed to be the state to which the input pins drive it, regardless of its proscribed state. Proscribed nodes in macros saved to disk are assigned the default initial state.

Quadrant
Function of the View menu of the Designer Display screen. Shows an entire quadrant on the screen at one-fourth scale.

Quadrants
Function of the View menu of the Designer Display screen. Shows the entire drawing on the screen at one-eighth scale.

R=:
Rotation/reflection number, located in the Function Selector window; determines the orientation at which components are added or moved. [Clicking the displayed value will reset R=: to 0.] There are eight codes for R=:, which may be accessed at the keyboard as follows:

0 = Point the component right and don't reflect it.

1 = Point the component down and don't reflect it.

2 = Point the component left and don't reflect it.

3 = Point the component up and don't reflect it.

4 = Point the component right and reflect it.

5 = Point the component down and reflect it.

6 = Point the component left and reflect it.

7 = Point the component up and reflect it.

Random — Option of the Patterns menu of the Pattern editor; used to create a random pattern in the specified data channel position.

Retrieve run — Option of the Simulation submenu of the Simulator pull-down menu; used to recall waveforms from disk in lieu of a simulation. Not available in the student version of MICRO-LOGIC II.

Run Asymmetric, Short, Long, or Both delays (A,S,L,B) — The analysis limit that allows the system to assign delays to each gate.

Run Fixed or Calculated delays (F,C) — The analysis limit that determines whether gate delays will be calculated on the basis of number of other gates they drive or on their fixed library values.

Runtime menu — Area located at the top of the screen to the right of the pull-down menu of the Simulator display screen. The Runtime menu has four options: Pause, Expand, Shrink, and Esc.

S=: — Grid step number, located in the Function Selector window; displays the current number of grids used when the cursor is moved or the display is scrolled. Pressing [+] on the numeric keypad doubles the stepping. Pressing [−] on the numeric keypad forces the stepping to 1. [Clicking the mouse at a value will reset S=: to 1.]

Save drawing — Function of the File menu in the Designer Display screen. Saves the current drawing to disk. Pressing [Esc] aborts the save.

Save pattern — Option of the Patterns menu of the Pattern editor; used to save the current pattern to disk.

Save run — Option of the Simulation submenu of the Simulator pull-down menu; used to save waveforms for later retrieval. Not available in the student version of MICRO-LOGIC II.

Scale factor — The number of screen pixels drawn per time interval on a simulation run; specifying this limit will compress or expand the displayed waveforms.

Screen — Option of the Print Reports function; used to send the report to the CRT screen only.

[Scroll control box] — The large box with a small circle in it, located at the top of the Designer Display screen to the right of the pull-down menu. The scroll control box controls display scrolling.

Scrolling — If [ScrollLock] is on, the following three actions can be accomplished: (1) Cursor keys will scroll the drawing in the direction specified; using the diagonal keys will scroll the screen diagonally. [(2) The mouse will scroll the drawing in the direction of the boundary at which the mouse is currently located.] (3) [Mouse or] [↑]/[↓] keys in the File dialog box will scroll the window of filenames. If [ScrollLock] is off, the following three actions can be accomplished: First, Cursor keys move the cursor in the direction specified. Second, [PgUp] moves the drawing view up one page; [PgDn] moves the drawing view down one page; [Home] moves the drawing view left one page; [End] moves the drawing view right one page. Third, [mouse or] [↑]/[↓] keys in the File selector box will select the next filename in the list.

Search — Function of the View menu of the Designer Display screen. Allows the user to search for a selected gate name or text fragment and to display the portion of the drawing at which the first instance of the chosen text appears. The user may search using partial or whole names and text. Pressing the F3 key will cause the system to search for the next instance of the chosen text.

Select — Option of the Component editor pull-down menu; used to access the Component library, which displays the entire list of components.

Set — Option of the Patterns menu of the Pattern editor; used to set or reset a specified set of pages (16 clock cycles each) in a specified block of data channels to 1 or 0.

Shape editor — A sophisticated shape-editing system activated by the Shapes option of the Editors menu of the Designer Display screen. Not available in the student version of MICRO-LOGIC II.

Shape names display Area to the bottom right of the Component editor display screen, used to list the set of shape names currently defined in the Shape library.

Shrink Option of the Runtime menu of the Simulator Display screen; used to halve the simulation scale factor, thereby compacting waveforms.

Simulate Menu located in the Designer Display screen; used to run a simulation.

Simulation Submenu of the Simulate pull-down menu; contains the following options: Home:Simulate, Edit analysis values, Edit nodes monitored, Patterns, Clocks, Save Run, Retrieve Run, and Text file.

Simulator The program that drives the network generated by the drawing module, complete with data channel patterns and clocks, and that produces the output waveforms.

[Step] Action located in the Function Selector window. Repetitively copies the circuitry enclosed in the currently defined box region. The user specifies the number of times to copy box contents and whether they should be copied in a vertical or horizontal direction, or both.

Text (1) The field that holds the text to be printed with the component; used mainly by macros to place identifying text on the DIP or FF shape employed. (2) Element located in the Function Selector window; refers to text placed at any of the grid locations in the drawing.

Text file Option of the Simulation submenu of the Simulator pull-down menu; used to generate an ASCII text file showing the states of each monitored node at the end of each clock cycle. Used primarily to generate test vectors. The filename is CIRCUITNAME.VEC. It can be used by most word processing systems.

Text X location The field that defines the x location of the optional text field in a component graphics symbol.

Text Y location The field that defines the y location of the optional text field in a component graphics symbol.

Tie A piece of text placed on two or more lines to connect them. A tie is defined with the reserved character /, placed at the beginning of the tie label. Where applied, a tie connects lines at the points where the labels appear.

Track lines Option of the Print Reports function; used to monitor line creations during the drawing-to-netlist conversion process; helpful in identifying faulty lines.

Unconnected pins report Option of the Print Reports function; used to list any component that has one or more input pins unconnected. Used primarily to spot faulty or incomplete wiring.

Undo Function located in the Function Selector that reverses the user's previous action; will also relocate a line to its original position. Undo cannot reverse the user's previous action if the word Undo is shaded.

Update Macros Function of the File menu of the Designer Display screen. Reads each macro drawing file from disk, creates its netlist, and saves it on the disk, attaching the extension .MAC to the filename. This function is used twice: first to generate netlist files for all drawing files, and second to save changes on a lower-level macro that appears in other macros by executing the changes wherever that macro appears.

View Menu located in the Designer Display screen. Used to perform the following view options: Capacity used, Quadrant, Quadrants, Comp text, Grid text, Comp numbers, and Search.

Waveform window The large central area of the Simulation Display screen that shows the waveforms as the simulation proceeds; window contents are scrolled automatically as the simulation progresses.

Wired-NOR Function of the Options menu of the Designer Display screen. Defaults the system to Wired-NOR, allowing the user to select the default gate type to be used when two or more component outputs are wired together.

Wired-OR Function of the Options menu of the Designer Display screen. Defaults the system to Wired-OR, allowing the user to select the default gate type to be used when two or more component outputs are wired together.

Zap Action located in the Function Selector. Deletes a component, line segment, grid text, [or box contents,] depending upon what the user has currently selected. [Lines that cross box boundaries are deleted up to the point where boundary and line meet. Elements that originate within a currently defined box are deleted when the user clicks the mouse.]

μ Menu located in the pull-down menu of all modules, used to access the title screen and to exit the current active module.

Index